Essays in Biochemistry

Other recent titles in the Essays in Biochemistry series:

Essays in Biochemistry volume 39: Programmed Cell Death
edited by T.G. Cotter
2003 ISBN 1 85578 148 4

Essays in Biochemistry volume 38: Proteases in Biology and Medicine
edited by N.M. Hooper
2002 ISBN 1 85578 147 6

Essays in Biochemistry volume 37: Regulation of Gene Expression
edited by K.E. Chapman and S. J. Higgins
2001 ISBN 1 85578 138 7

Essays in Biochemistry volume 36: Molecular Trafficking
edited by P. Bernstein
2000 ISBN 1 85578 131 X

Essays in Biochemistry volume 35: Molecular Motors
edited by G. Banting and S.J. Higgins
2000 ISBN 1 85578 103 4

Essays in Biochemistry volume 34: Metalloproteins
edited by D.P. Ballou
1999 ISBN 1 85578 106 9

Essays in Biochemistry volume 33: Molecular Biology of the Brain
edited by S.J. Higgins
1998 ISBN 1 85578 086 0

Essays in Biochemistry volume 32: Cell Signalling
edited by D. Bowles
1997 ISBN 1 85578 071 2

volume 40 2004

Essays in Biochemistry

The Nuclear Receptor Superfamily

Edited by I.J. McEwan

Portland Press

Essays in Biochemistry is published by Portland Press Ltd
on behalf of the Biochemical Society

Portland Press
59 Portland Place
London W1B 1QW, U.K.
Fax: 020 7323 1136;
e-mail: editorial@portlandpress.com
www.portlandpress.com

British Library Cataloguing-in-Publication Data
A catalogue record for this book is available from the British Library

ISBN 1 85578 150 6
ISSN 0071 1365 (print)
ISSN 1744 1358 (online)

Typeset by Portland Press Ltd
Printed in Great Britain by Information Press Ltd, Oxford

Contents

Preface ....

Authors ....

Abbreviations ....

1 **Sex, drugs and gene expression: signalling by members of the nuclear receptor superfamily**
Iain J. McEwan

Abstract 1
Introduction 1
"*Evolution... is a change from an indefinite, incoherent homogeneity, to a defined coherent heterogeneity*" Herbert Spenser (*First Principles*) 2
"*What is the character of a family to a hypothesis?*"
Laurence Sterne (*Tristram Shandy*) 2
"*Action is transitory, — a step, a blow...*"
William Wordsworth (*The Borders*) 6
"*Accidents will occur in the best-regulated families*"
Charles Dickens *(David Copperfield)* 7
Conclusions and future perspectives 8
Summary 9
References 9

2 **The evolution of the nuclear receptor superfamily**
Héctor Escriva, Stéphanie Bertrand and Vincent Laudet

Abstract 11
Introduction 11
Appearance and diversification of the NR superfamily 13
Evolution of the NRs functional characteristics 17
Comparative genomics of NRs 20
Conclusions and perspectives 22
Summary 23
References 24

3 Overview of the structural basis for transcription regulation by nuclear hormone receptors

Raj Kumar, Betty H. Johnson, and E. Brad Thompson

Abstract 27
Introduction 28
NTD 30
DBD 33
LBD 34
NHRs and co-regulatory proteins 36
Conclusions 37
Summary 37
References 38

4 Role of molecular chaperones in steroid receptor action

William B. Pratt, Mario D. Galigniana, Yoshihiro Morishima and Patrick J. M. Murphy

Abstract 41
Introduction 41
Hsp90 acts on the ligand-binding domain (LBD) of steroid receptors 42
Assembly of receptor–hsp90 heterocomplexes 44
Mechanism of cleft opening 47
Hsp90 and GR function *in vivo* 50
Conclusion 56
Summary 56
References 57

5 DNA recognition by nuclear receptors

Frank Claessens and Daniel T. Gewirth

Abstract 59
Introduction 60
The HREs 61
The DBDs 63
Discrimination of the core hexamers 63
Dimerization controls specificity 66
Exceptions to the rules 67
Allosteric effects of HREs 68
Other functions of the DBD 69
General conclusions 69
Summary 70
References 70

6 Transcriptional activation by nuclear receptors

Mari Luz Acevedo and W. Lee Kraus

Abstract 73
Introduction 73
Chromatin: the physiological template for NR-dependent transcription74
The RNA pol II transcriptional machinery 74
NR co-activators: distinct classes with multiple activities 76
Histone modifications and the histone code 80
Cofactor recruitment and activity at NR-regulated promoters....81
Shutting off NR-dependent transcription: factor modification, recycling, redistribution and turnover 84
Cell-type-specificity and co-integration of multiple signalling pathways..85
Conclusion 86
Summary....86
References....87

7 Gene repression by nuclear hormone receptors

Udo Moehren, Maren Eckey and Aria Baniahmad

Abstract 89
Introduction....90
Gene-silencing by co-repressors 90
NHR action on chromatin....94
Gene silencing by NHRs through interference with the basal transcription machinery....96
Hormone receptor inactivation by anti-hormones....96
Negative response elements....97
Transcriptional repression by NHR through inhibition of signalling pathways via crosstalk....98
Impact of gene repression by NHRs on development and disease....100
Conclusions 101
Summary....101
References 101

8 Receptor mechanisms of rapid extranuclear signalling initiated by steroid hormones

Viroj Boonyaratanakornkit and Dean P. Edwards

Abstract....105
Introduction....106
Extranuclear signalling actions of progesterone: classical intracellular progesterone receptor (PR)....107

Extranuclear progesterone signalling: role of a novel mPR......110
Extranuclear signalling of oestrogen: role of the classical ER......112
Extranuclear signalling actions of oestrogen: role of novel membrane receptors?......113
Cell-membrane localization of a subpopulation of classical ERs......114
The role of extranuclear ER signalling *in vivo*......115
Conclusions......116
Summary......117
References......118

9 Nuclear receptors and disease: androgen receptor
Bruce Gottlieb, Lenore K. Beitel, Jianhui Wu, Youssef A. Elhaji and Mark Trifiro

Abstract......121
Introduction......122
Structure–function relationships of the AR protein and gene......122
Diseases as a result of mutations in the AR gene......125
Diseases directly associated with *AR* CAG repeat-length variation: spinobulbar muscular atrophy (SBMA; Kennedy disease)......130
AR CAG tract-length variation as a risk factor for disease......131
Conclusions......134
Summary......134
References......135

10 Glucocorticoid and mineralocorticoid receptors and associated diseases
Tomoshige Kino and George P. Chrousos

Abstract......137
Introduction......138
Structure and actions of GRs and MRs......139
Natural physiological steroid hormone 'resistance' in animals......143
Pathological changes of GR and MR activities in humans......144
Conclusions......152
Summary......153
References......153

11 Nuclear receptors in disease: the oestrogen receptors

Maria Nilsson, Karin Dahlman-Wright and Jan-Åke Gustafsson

Abstract ... 157
The oestrogen receptors (ERs) ... 158
ER modulators in physiology and disease ... 159
ERs and breast cancer ... 159
ERs and prostate cancer ... 160
ERs and cardiovascular disease ... 161
ERs and osteoporosis ... 161
ERs and diseases of the central nervous system (CNS) ... 162
ERs and diseases of the immune system ... 162
ER polymorphisms and mutations in relation to disease ... 162
Conclusion ... 164
Summary ... 165
References ... 165

12 Nuclear receptors and human disease: thyroid receptor β, peroxisome-proliferator-activated receptor γ and orphan receptors

Mark Gurnell and V. Krishna K. Chatterjee

Abstract ... 169
Introduction ... 170
TRβ and the syndrome of resistance to thyroid hormone (RTH) ... 173
PPARγ and the syndrome of PPARγ ligand resistance (PLR) ... 177
HNF4α and MODY1 ... 181
SHP and obesity ... 182
SF1 and DAX1 in disorders of gonadal and adrenal development ... 183
PNR and enhanced S-cone syndrome (ESCS) ... 184
NURR1 and abnormalities of dopaminergic neurotransmission: familial Parkinson's disease and schizophrenia/manic-depressive disorder ... 185
Conclusions ... 186
Summary ... 186
References ... 187

Subject index ... 191

Preface

Communication between cells is of central importance in multicellular organisms. Indeed, it could be argued that it is one of the defining features of multicellularity, as it allows for cell specialization and cell organization into tissues in a controlled and regulated manner. Nuclear receptors are a large family of signalling proteins that play crucial roles during embryonic development and in the regulation of reproductive and metabolic functions in the adult.

Nuclear receptors share a common architecture at the protein level, but a remarkable diversity is observed in terms of natural ligands and xenobiotics that bind to and regulate receptor function. The defining feature of the ligands is that they are small, lipophilic molecules that are thought to diffuse across the plasma membrane. Nuclear receptors act by regulating the patterns of gene expression in target cells and can be thought of as 'ligand-activated' transcription factors. However, a significant proportion of family members have been described as 'orphans' as the natural ligand, if it exists, remains to be identified. This volume of *Essays in Biochemistry* focuses on recent exciting developments in nuclear receptor action, as well as highlighting future research goals. It is timely, as it is almost 20 years since the first steroid receptor cDNAs were cloned, and over 45 years since the first clues were uncovered as to the molecular actions of what turned out to be a large family of diverse signalling proteins.

After a brief overview of the area, the volume starts with a discussion on the evolution of the functional diversity that is found within the nuclear receptor superfamily by Laudet and colleagues. These authors also consider what information can be gained from comparative genome-wide analysis. There then follows two essays that examine the structure–function relationships, by Thompson and colleagues, and by Claessens and Gewirth. The latter authors focus on the DNA-binding properties of family members. The distinct roles that protein–protein interactions play in nuclear receptor action are considered by essays on molecular chaperone complexes (Pratt and colleagues), co-activator complexes (Acevedo and Kraus), co-repressor complexes (Baniahmad and colleagues) and the extranuclear actions of progesterone and oestradiol receptors (Boonyaratanakornkit and Edwards). The preceding discussions on the structure and biochemical properties of nuclear receptors lead into the final four essays on the androgen receptor (Trifiro and colleagues), glucocorticoid and mineralocorticoid receptors (Kino and Chrousos), oestrogen receptors (Gustafsson and colleagues), and thyroid hormone receptors, peroxisome-proliferator-activated receptors and orphan receptors (Gurnell and Chatterjee). These essays address the question of the physiological roles of different nuclear receptors and how genetic alterations in receptor proteins disrupts signalling and underpins a diverse range of pathological conditions, including cancers, reproductive and developmental defects, and metabolic disorders, such as diabetes and obesity.

I would like to express my sincere thanks to the authors for their excellent contributions and discussing their individual areas of expertise in a clear and erudite manner. I am also grateful to the many reviewers for their constructive comments and suggestions on the submitted manuscripts. It was my aim at the beginning of this enterprise that this volume would be accessible and of benefit to both undergraduate and postgraduate students, as well as their teachers, in the biological and medical science disciplines, who wish to learn about nuclear receptors. Now that it is complete, I hope those already working in nuclear receptor research will also find the topics covered of interest and stimulating. My thanks also go to Mike Cunningham and Portland Press Ltd for their hard work and diligence in ensuring the high quality of this book.

Iain J. McEwan
Aberdeen, UK
April 2004

Authors

Iain McEwan's scientific career began in Glasgow as an undergraduate student at the University of Strathclyde, graduating in 1983 with a first in Biochemistry (B.Sc.), before obtaining a Ph.D. degree from the University of Glasgow in 1997. He has worked as a research fellow at the Friedrich Miescher Institute in Basel, Switzerland, and the Karolinska Institute in Stockholm, Sweden. He was appointed to a lectureship at the University of Aberdeen in 1997, where he is currently a Senior Lecturer in the School of Medical Sciences. His research focuses on the molecular mechanisms of action of the androgen receptor in health and disease, and his group is particularly interested in understanding the dynamic relationships between receptor conformation and interactions with protein-binding partners and DNA-response elements.

Hector Escriva studied biology at the Universitat de Valencia, Spain, and then moved to the Institut Pasteur de Lille, France, where he undertook a Ph.D. in molecular parasitology, entitled 'Nuclear receptors in *Schistosoma mansoni*: molecular cloning and characterization of an RXR homolog. Molecular phylogeny of the nuclear receptor superfamily'. At present he is working on the evolution of the nuclear receptor superamily as a CNRS Research Scientist at the Ecole Normale Supérieure de Lyon, France. **Stéphanie Bertrand** studied biology at the Université de Lyon, France. She worked towards a Masters thesis on amphibian metamorphosis and she is now working on her Ph.D. at the Ecole Normale Supérieure de Lyon, on the comparative genomics of the nuclear hormone receptor superfamily. **Vincent Laudet** studied biochemistry at the Université Louis Pasteur de Strasbourg, France. He did a Ph.D. in the Laboratoire d'Oncologie Moléculaire at the Institut Pasteur de Lille, France, on the structure and evolution of the nuclear hormone receptors. After his Ph.D., he worked for several years as a CNRS Research Scientist at the Institut Pasteur de Lille, and then he moved to the Ecole Normale Supérieure de Lyon as a Professor and where he is now the group leader of the Structure and Evolution of the Nuclear Hormone Receptors Group.

Raj Kumar is Assistant Professor of Human Biological Chemistry and Genetics at the University of Texas Medical Branch, Galveston, TX. He graduated from the University of Lucknow, India, in 1987 with a B.Sc. in physics, chemistry and mathematics, and was awarded an M.Sc. (chemistry in 1989) and a Ph.D. (chemistry in 1995) from the same institution. **Betty H. Johnson** graduated from Southwestern University, Georgetown, TX, with a B.S. in chemistry in 1955. After several research and teaching positions, she received an

M.S. degree in biochemistry in 1983 from University of Texas Medical Branch, where she is currently a Research Scientist. **E. Brad Thompson** received a B.A. in premedicine from Rice University, Houston, TX, in 1955. This was followed by a period of study at the University of Cambridge (biochemistry) and then at Harvard Medical School (medicine), from which he received an M.D. (1960). After a medical residency at the Columbia Presbyterian Medical Center, New York, and research work at the National Institutes of Health, Bethesda, MD, he moved to the University of Texas Medical Branch, where he is Professor of Human Genetics. His research interests include the actions of glucocorticoids on leukaemic cells and structural studies of the glucocorticoid receptor.

William Pratt, M.D., is Professor of Pharmacology at the University of Michigan, Ann Arbor, MI. He has worked in the field of steroid-hormone action for 40 years, and together with Etienne Baulieu and David Toft, he discovered the interaction of hsp90 with steroid receptors. **Mario Galigniana**, Ph.D., is an Assistant Research Scientist at the University of Michigan. He has worked on mineralocorticoid and glucocorticoid receptors for 15 years, and over the past 9 years, he has determined roles for hsp90 and hsp90-bound immunophilins in glucocorticoid receptor translocations from the cytoplasm to the nucleus. **Yoshihiro Morishima**, M.D., Ph.D., is a Postdoctoral Researcher at the University of Michigan. He has spent the last 6 years working on the mechanism of assembly of glucocorticoid receptor–hsp90 heterocomplexes. **Patrick Murphy**, Ph.D., is a Postdoctoral Researcher at the University of Michigan. For the past 5 years, he has worked on the stoichiometry and structure of the hsp90/hsp70-based chaperone machinery.

Frank Claessens obtained a Masters degree in biology (1983), and a Masters degree in medical biotechnology (1984) at the University of Leuven, Belgium; he received his Ph.D. in 1989. He did postdoctoral work at the Imperial Cancer Research Fund, London, before becoming a faculty member at the Medical Faculty of the University of Leuven. **Daniel Gewirth** received his B.S. in chemistry from the University of Chicago in 1982, and his Ph.D. in molecular biophysics and biochemistry from Yale University. He did postdoctoral work at Harvard and Yale Universities, and joined the faculty of Duke University in 1998.

Mari Luz Acevedo graduated from the University of Puerto Rico, Rio Piedras, with a B.S. She is a senior doctoral candidate in the Graduate Field of Biochemistry, Molecular and Cell Biology at Cornell University, Ithaca, NY, and is minoring in pharmacology. Her thesis project is an investigation of the role of co-activators in transcriptional regulation by the oestrogen receptor. Her graduate studies are funded by the National Institute of Diabetes and Digestive and Kidney Diseases (NIDDK), a part of the National Institutes of Health. **W. Lee Kraus** obtained an M.S. and a Ph.D. from the University of Illinois, Champaign-Urbana, IL. He has been an Assistant Professor in the

Department of Molecular Biology and Genetics at Cornell University, Ithaca, NY, since 1999, and an Adjunct Assistant Professor of Pharmacology at the Weill Medical College of Cornell University, New York since 2000. His research interests include the molecular biology of nuclear-receptor-mediated signalling and the biochemistry of transcriptional regulation with chromatin templates. His teaching interests include the molecular basis of human disease. The work in his laboratory is supported by the Burroughs Wellcome Fund, the NIDDK, the American Cancer Society and the Komen Breast Cancer Foundation.

Udo Moehren obtained his Master of Science degree in biology at the Justus-Liebig University, Giessen, Germany, where he is currently working as a postgraduate student in Aria Baniahmad's group. His research interests are the molecular mechanisms of anti-hormone action on the androgen receptor and the involvement of co-repressors. **Maren Eckey** studied biology at the Justus-Liebig University, and completed her degree with a laboratory diploma thesis (Masters of Sciences) based on the analysis of gene silencing by thyroid hormone receptors and the co-repressor Alien. As a graduate student, she continued to work in this field with Aria Baniahmad, analysing novel repression mechanisms mediated by the co-repressor Alien on the chromatin level. **Aria Baniahmad** studied biology at the University of Würzburg and the University of Hohenheim, Germany. He was awarded a Ph.D. at the Max-Planck-Institute for Biochemistry in Munich, Germany, for identifying the repressive regulation by thyroid hormone and retinoic acid receptors. He the joined Bert O'Malley's lab at Baylor College of Medicine, Houston, TX, U.S.A. After moving back to Germany, he became independent group leader and principal investigator at the Justus-Liebig University, whose work analyses gene-silencing mechanisms by nuclear hormone receptors.

Viroj Boonyaratanakornkit received a degree in Medical Technology from Chulalongkorn University, Bangkok, Thailand, in 1988 and a Ph.D. from Loma Linda University, CA, U.S.A., in 1996. He then moved to the University of Colorado Health Sciences Center, Denver, CO, as a post-doctoral fellow and is currently an Instructor in the Department of Pathology. **Dean P. Edwards** is graduated with a B.S. in zoology from Ohio University, Athens, OH, in 1969. He received a Ph.D. from the Medical College of Georgia, Augusta, GA, in 1976 and followed this as a research fellow at the University of Texas, San Antonio, TX, working on steroid receptors and breast cancer. He is currently a Professor at the Department of Pathology, in the School of Medicine at the University of Colorado Health Sciences Center, Denver, CO, where his research continues to focus on steroid receptors and hormone action in breast cancer.

Bruce Gottlieb, Ph.D., is a geneticist/molecular biologist who is the curator of the Androgen Receptor Gene Mutations Database and is particularly interested in the development of gene-mutation databases and the utilization of computer modelling to relate protein structure to function. He is a project director at the Lady Davis Institute for Medical Research, Montréal, Canada, an Adjunct Professor in the Faculty of Medicine at McGill University, Montréal, Canada, and a Professor of Biology at John Abbott College, Ste. Anne de Bellevue, Canada. **Lenore K. Beitel**, Ph.D., is a biochemist/molecular biologist who is interested in many aspects of the structure–function relationship of the androgen receptor and, in particular, with androgen-receptor-interacting proteins, and spinobulbar muscular atrophy. She is a research scientist at the Lady Davis Institute for Medical Research and an Assistant Professor in the Faculty of Medicine at McGill University. **Janbui Wu**, Ph.D., is a chemist that is interested in using computer-generated molecular dynamic modelling to predict the effect of mutations on protein structure and function. He is a project director at the Lady Davis Institute for Medical Research and an Assistant Professor in the Department of Oncology at McGill University. **Yossef Elhaji**, M.Sc., is a Ph.D. candidate in the Department of Human Genetics at McGill University. **Mark Trifiro**, M.D., is a clinical and research molecular endocrinologist, and heads the Molecular Endocrinology Laboratory at the Lady Davis Institute for Medical Research. He is an Associate Professor of Medicine at McGill University.

George P. Chrousos is the Chief of the Pediatric and Reproductive Endocrinology Branch at the U.S. National Institute of Child Health and Human Development (NICHD), the National Institutes of Health (NIH), Bethesda, MD, Clinical Professor of Pediatrics and Physiology at Georgetown University Medical School, Washington, DC, and Professor of Pediatrics, Athens University Medical School, Athens, Greece. **Tomoshige Kino** is a Staff Scientist of the Pediatric and Reproductive Endocrinology Branch at the NICHD.

Maria Nilsson, M.Sc., received her degree from the University of Umeå, Sweden. She is currently working on nuclear receptor signalling in physiology and disease as a Ph.D. student in Professor Gustafsson's laboratory at the Department of Biosciences, Karolinska Institute, Huddinge, Sweden. **Karin Dahlman-Wright** received her Ph.D. from the Karolinska Institute in 1991. After a post-doctoral period, she joined Pharmacia and Upjohn in 1995, where she held different line- and project-management positions. Since 2000, she has been a group leader at the Department of Biosciences at the Karolinska Institute, where her major focus is to apply functional genomics approaches to study the role of oestrogen hormone and its receptors in physiology and disease. **Jan-Åke Gustafsson**, M.D. Ph.D., is Professor of Medical Nutrition and Director of the Center for Biotechnology at Novum, the South Campus of the Karolinska Institute; his main interest is nuclear receptor signalling. He is a foreign associate of the U.S. National Academy of Sciences.

Mark Gurnell trained as an undergraduate at St Bartholomew's Hospital, London, and it was there that he first developed his interest in endocrinology under the combined tutorships of Professors G.M. Besser, J.A.H. Wass and A.B. Grossman. His interest in the molecular basis of endocrine disorders subsequently took him to St John's College, Cambridge as a Wellcome Training Fellow under the supervision of Professor V.K.K. Chatterjee, where he completed his Ph.D. entitled 'The roles of the human thyroid hormone β receptor and the peroxisome-proliferator-activated receptor γ (PPARγ) in human disease'. Having completed his clinical training in Cambridge, he has continued to work in close collaboration with Professor Chatterjee, exploring the roles of orphan nuclear receptors in human metabolism. **V. Krishna K. Chatterjee** was an undergraduate at Cambridge and completed his clinical training in Oxford. He trained as an endocrinologist at the Hammersmith Hospital, London and subsequently undertook research in the Thyroid Unit at Massachusetts General Hospital, Boston, MA, with Professor Larry Jameson. In 1990, he returned to the Department of Medicine at Cambridge as a Wellcome Senior Clinical Research Fellow and was appointed Professor of Endocrinology there in 1998. His research interests include the syndrome of resistance to thyroid hormone and defects in nuclear hormone receptors (including PPARγ) in human disorders. He is also evaluating the role of DHEA (dehydroepiandrosterone) hormone replacement in adrenal insufficiency.

Abbreviations

AD	activation domain
AF	activation function
Aha	activator of heat-shock protein 90 ATPase
AHC	adrenal hypoplasia congenita
AIS	androgen-insensitivity syndrome
AP-1	activator protein 1
APML	acute promyelocytic leukaemia
AR	androgen receptor
ARKO	aromatase-knockout
ASSC	amiloride-sensitive sodium channel
AVP	arginine vasopressin
BAG-1	Bcl-2-associated gene product-1
BMD	bone mineral density
Brg1	brahma-related gene-1
CAIS	complete androgen-insensitivity syndrome
CaP	prostate cancer
CAR	constitutive androstane receptor
CARM	co-activator-associated arginine methyltransferase
CBP	cAMP-response-element-binding protein-binding protein
CHIP	C-terminus of heat-shock-protein-70-interacting protein
ChIP	chromatin immunoprecipitation
CHO	Chinese-hamster ovary cells
CNS	central nervous system
COUP-TF	chicken ovalbumin upstream promoter-transcription factor 1
CRC	chromatin-remodelling complex
CREB	cAMP-response-element-binding protein
CRH	corticotropin-releasing hormone
CTE	C-terminal extension
CVD	cardiovascular disease
CyP	cyclophilin
DAX1	dosage-sensitive sex reversal-adrenal hypoplasia congenita critical region on the X chromosome gene 1
DBD	DNA-binding domain

D-box	distal box
DR	direct repeat
EcR	ecdysone receptor
EGF	epidermal growth factor
eNOS	endothelial nitric oxide synthase
ER	oestrogen receptor
ERK	extracellular-signal-regulated kinase
ERKO	oestrogen-receptor-knockout
ERR	oestrogen-related receptor
ESCS	enhanced S-cone syndrome
FKBP	FK506-binding protein
FTZ-F1	fushi tarazu factor 1
FXR	farnesoid X receptor
GCNF	germ cell nuclear factor
GPCR	G-protein-coupled receptor
GR	glucocorticoid receptor
GRE	glucocorticoid-response element
GRTH	generalized resistance to thyroid hormone
HAT	histone acetyltransferase
HDAC	histone deacetylase
Hip	heat-shock-protein-70-interacting protein
HNF	hepatocyte nuclear factor
Hop	heat-shock protein organizer protein
HP1	heterochromatin-associated protein 1
HPA axis	hypothalamic/pituitary/adrenal axis
HPG axis	hypothalamic/pituitary/thyroid axis
Hr	Hairless
HRE	hormone-response element
hsp	heat-shock protein
ID	interaction domain
IR	inverted repeat
LBD	ligand-binding domain
MAPK	mitogen-activated protein kinase
MNAR	modulator of non-genomic activity of oestrogen receptor
MODY1	maturity onset diabetes of the young type 1
mPR	membrane progesterone receptor
MR	mineralocorticoid receptor
NCoR	nuclear receptor co-repressor
NGFI-B	nerve growth factor inducible factor I-B
NHR	nuclear hormone receptor
nHRE	negative hormone-response element
NR	nuclear receptor

NTD	N-terminal domain
NURD	nucleosome remodelling and histone deacetylation
NURR1	NUR-related factor 1
PAIS	partial androgen-insensitivity syndrome
P-box	proximal box
pCAF	p300/CBP-associated factor
PHA1	pseudohypoaldosteronism type 1
PI 3-kinase	phosphoinositide 3-kinase
PIC	pre-initiation complex
PLR	peroxisome-proliferator-activated receptor γ ligand resistance
PNR	photoreceptor-specific nuclear receptor
PPAR	peroxisome-proliferator-activated receptor
PPIase	peptidylprolyl isomerase
PPP1R3A	protein phosphatase 1 regulatory subunit 3A
PR	progesterone receptor
PRMT	protein arginine methyltransferase
PRTH	pituitary resistance to thyroid hormone
RAR	retinoic acid receptor
RIP140	receptor-interacting protein 140
RNA pol II	RNA polymerase II
ROR	retinoid-related orphan receptor
RTH	resistance to thyroid hormone
RXR	retenoid X receptor (9-*cis*-retinoic acid receptor)
SAP	Sin3A–Sin-associated protein
SBMA	spinobulbar muscular atrophy
SERM	selective oestrogen receptor modulator
SF1	steroidogenic factor 1
SH domain	Src homology domain
SHBG	sex-hormone-binding globulin
SHP	small heterodimer partner
SMRT	silencing mediator for retinoic acid receptor and thyroid hormone receptor
SMRTER	silencing mediator of repressed transcription
SNP	single-nucleotide polymorphism
SRC	steroid receptor co-activator
STAT	signal transducer and activator of transcription
T2DM	Type II diabetes mellitus
TAF	TATA-box-binding-protein-associated factor
TBP	TATA-box-binding protein
TF	transcription factor
TMAO	trimethylamine *N*-oxide
TPR	tetratricopeptide repeat

TR	thyroid hormone receptor
TZD	thiazolidinedione
USP	ultraspiracle
VDR	vitamin D receptor
WT	wild-type
XPR	*Xenopus* homologue of mammalian nuclear progesterone receptor

1

Sex, drugs and gene expression: signalling by members of the nuclear receptor superfamily

Iain J. McEwan[1]

School of Medical Sciences, IMS Building, University of Aberdeen, Foresterhill, Aberdeen AB25 2ZD, Scotland, U.K.

Abstract

It is almost 20 years since the first steroid receptor cDNAs were cloned, a development that led to the concept of a superfamily of ligand-activated transcription factors: the nuclear receptors. Natural ligands for nuclear receptors are generally lipophilic in nature and include steroid hormones, bile acids, fatty acids, thyroid hormones, certain vitamins and prostaglandins. Nuclear receptors act principally to directly control patterns of gene expression and play vital roles during development and in the regulation of metabolic and reproductive functions in the adult organism. Since the original cloning experiments, considerable progress has been made in our understanding of the structure, mechanisms of action and biology of this important family of proteins.

Introduction

The physiological actions of steroid hormones have been recognized for more than 100 years. The role of the testes in producing a 'male hormone' was demonstrated experimentally by the pioneering experiments of A.A.

[1]*E-mail iain.mcewan@abdn.ac.uk*

Berthold in 1849 [1], while in 1896, G. Beatson showed that removal of the ovaries led to regression of breast tumours [2]. It was then only a matter of time before the molecular structures of these versatile messengers were determined and they were chemically synthesized. The first breakthrough in unravelling the molecular mechanism of action of steroid hormones was undoubtedly the availability of radiolabelled steroid molecules that allowed the identification of receptor proteins within target cells and led to the formulation of the two-step model for steroid action by Jensen and colleagues (reviewed in [2]). Some 20 years later, in the mid-1980s, the next dramatic development took place with the cloning of the cDNAs for the glucocorticoid receptor (GR), the oestrogen receptor (ER) and the progesterone receptor (PR). This rapidly led to the cloning of receptor proteins for non-steroid ligands, such as thyroid hormones (TRs) and retinoic acid (RAR), and the nuclear receptor superfamily was born ([3] and references therein).

"Evolution... is a change from an indefinite, incoherent homogeneity, to a defined coherent heterogeneity" Herbert Spenser (*First Principles*)

The nuclear receptor superfamily represents a disparate collection of proteins that share a common structural organization. These proteins in many, if not all, cases transduce signals that are important for animal development, metabolic regulation and reproduction. Representatives of the family have been identified in all classes of metazoans, but, strikingly, appear to be absent in yeasts and plants (see Chapter 2). A comparison of amino-acid sequences of the DNA- and ligand-binding domains has led to the classification of six subfamilies, and compelling evidence that the diversity within the family arose from two separate gene duplication and diversification events. It also seems likely that the ancestral protein was an orphan receptor and that ligand binding was an acquired property that occurred more than once independently (see Chapter 2). The availability of complete genome sequences for a number of metazoans has revealed some interesting observations regarding the occurrence of nuclear receptors in different species. The human genome sequence reveals 48 members of the family, with just under half (21 genes) representing receptors with known ligands. *Drosophila* contains 21 family members, while the nematode worm *Caenorhabditis elegans* has more than 200 nuclear receptor genes, but lacks any of the known ligand-binding members of the family.

"What is the character of a family to a hypothesis?" Laurence Sterne (*Tristram Shandy*)

Members of the nuclear receptor superfamily have a modular structure that consists of a C-terminal ligand-binding domain (LBD) linked by a hinge region to the DNA-binding domain (DBD), which is then flanked by a unique N-terminal domain (NTD) (Figure 1). The NTD is the most variable region in terms of amino-acid sequence and length, ranging from several hundred amino acid residues for certain steroid receptors to less than 100 for TRα, RAR and the vitamin D receptor (VDR). This region contains sequences [termed activation function 1 (AF1)] that are important for receptor-dependent transactivation. Structural analysis of the AF1 domain of several members of the family indicates that this region is structurally flexible and adopts a more folded conformation upon specific protein–protein interactions (see Chapter 3).

The folding of the core DBD is a defining feature of this family of transcription factors. The DBD is characterized by eight conserved cysteine residues that co-ordinate two zinc ions and consists of two α-helices that are folded perpendicular to each other (see Chapters 3 and 5). The first zinc module forms the 'recognition helix', which mediates specific amino-acid–base-pair interactions. In addition, a region flanking the core DBD, termed the 'C-terminal extension' (CTE) also plays a role in protein–DNA and/or protein–protein interactions. The CTE has been shown to form an α-helix for a number of non-steroid receptors [i.e. retinoid X receptor (RXR), TR and VDR], but appears to be disordered in steroid-binding members of the family (see Chapter 5). Members of the nuclear receptor superfamily recognize and bind DNA half-sites consisting of the sequences AGAACA or AGGTCA; the latter can be arranged as inverted repeats with 3-base-pair spacing, as direct repeats with variable spacing or as single elements. The different configurations of half-sites permit members of the family to bind as homodimers (i.e.

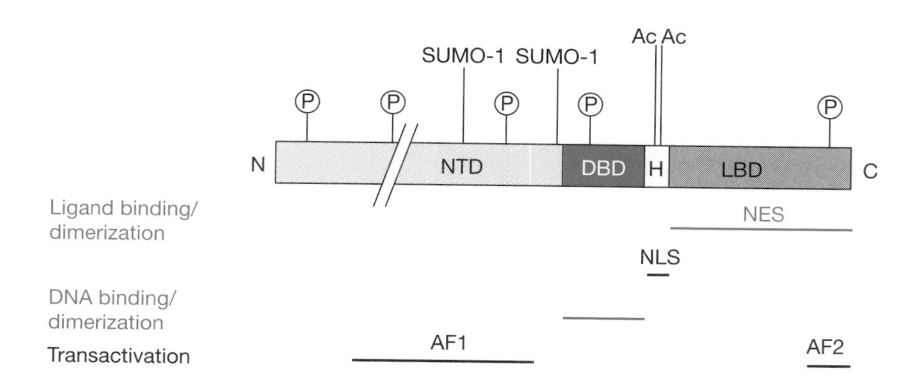

Figure 1. Schematic representation of the domain organization of members of the nuclear receptor superfamily
The LBD, DBD and the structurally distinct NTD are indicated. The positions of nuclear localization (NLS) and nuclear export (NES) signals are shown [15]. Regions of the receptor protein that are important for ligand binding, DNA binding, dimerization and transactivation are indicated below the protein. Target sites of potential post-translational modifications are indicated above the protein: Ac, acetylation; P, phosphorylation, and SUMO-1 (small ubiquitin-like modifier 1). H, hinge region.

steroid receptors), heterodimers with RXR [i.e. the constitutive androstane receptor β, the peroxisome-proliferator-activated receptor (PPAR), RAR, TR and VDR] or as monomers [i.e. NGFI-B (nerve growth factor inducible factor I-B), RORα (retinoid-related orphan receptor α) and SF1 (steroidogenic factor 1)] respectively (see Figure 2 and Chapter 5).

The common structural organization of nuclear receptors that is observed at the protein level belies the variety in structure of both natural ligands, including steroids, oxysterols, bile acids, thyroid hormones, retinoids and prostaglandins, as well as a plethora of drugs, that can bind and regulate receptor activity (Figure 3). However, are all members of the family actually receptor proteins? Clearly, some of members of the family are *bona fide* receptor proteins with identified and well-characterized natural ligands; however a significant number of family members have yet to have a ligand identified and have been termed 'orphan receptors', and, in certain cases, it seems likely that no ligand is required for protein function. In addition to ligand binding, the LBD also contains sequences that mediate protein–protein interactions, including dimerization, interactions with molecular chaperone complexes (see Chapter 4), and the binding of co-activators (see Chapters 3 and 6) or co-repressors (see Chapter 7).

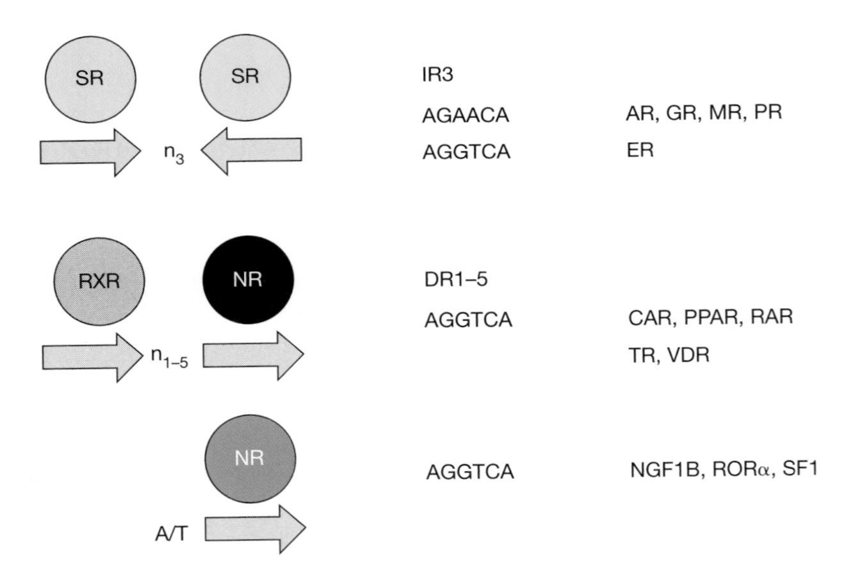

Figure 2. Representative members of the family and cognate DNA-response elements
Nuclear receptors recognize the half-site DNA sequences AGAACA or AGGTCA arranged as inverted repeats (IRs), direct repeats (DRs) or single sites flanked by an A/T sequence. Steroid receptors bind as homodimers to IRs, while obligate heterodimer partners of RXR bind to direct repeats with variable spacing. The orphan receptors, NGFI-B, SFI and RORα bind to DNA as monomers.

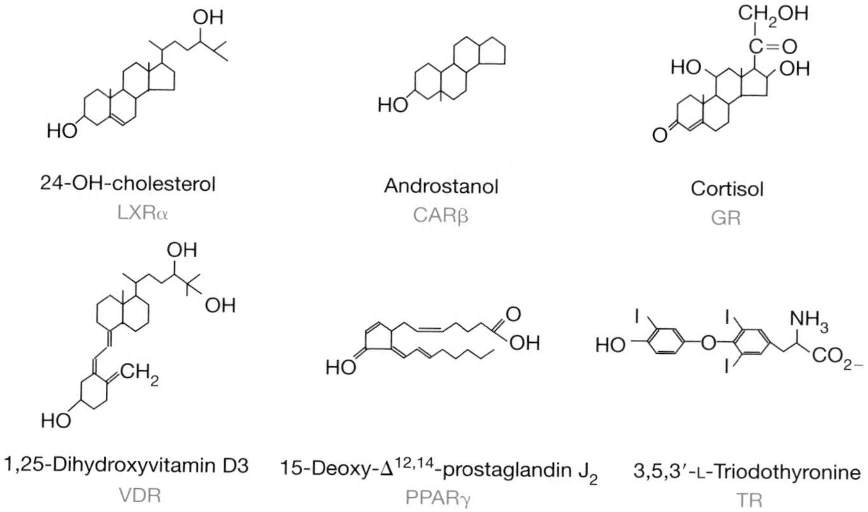

24-OH-cholesterol
LXRα

Androstanol
CARβ

Cortisol
GR

1,25-Dihydroxyvitamin D3
VDR

15-Deoxy-$\Delta^{12,14}$-prostaglandin J_2
PPARγ

3,5,3′-L-Triodothyronine
TR

Figure 3. A selection of natural ligands for members of the nuclear receptor superfamily with the receptor listed below
LXRα, liver X receptor; CARβ, constitutive androstane receptor.

The three-dimensional structures of the LBD of both ligand-activated and orphan receptors have been solved (see Chapter 3). The overall folding of this domain is remarkably conserved, but what is striking are the differences in the dimensions of the 'ligand-binding pocket' and the correlation with putative ligands. Thus, PPARγ has a relatively large binding pocket, consistent with the range of natural and xenobiotic ligands that can bind and activate this receptor. This contrasts with the binding pocket of other receptors with more restrictive ligand profiles. Of particular note is the structure of the LBD of the NGFI-B subfamily of orphan receptors, for which the putative ligand-binding pocket is occupied by the bulky ring side-chains of four phenylalanine residues, which appear to act as an 'internal ligand' (see Chapter 3).

Apart from the obvious regulatory action of natural ligands and drugs, how else might the functions of these proteins be controlled? It is now clear that the binding of nuclear receptors to specific DNA sequences within the genome may not simply function to tether the protein to promoter and/or enhancer regions. There is good evidence that DNA response elements can serve as allosteric regulators of receptor function (see Chapters 3 and 5). Another means of potential regulation that has generated considerable excitement is different forms of post-translational modification of receptor proteins, including phosphorylation, acetylation, sumoylation [modification with SUMO-1 (small ubiquitin-like modifier 1)] or ubiquitination (Figure 1). In many cases, the site, or sites, of modification has been identified, with serine and lysine residues within the NTD being phosphorylated and sumolyated respectively, and lysine residues within the hinge region of the AR and ER being acetylated (see Chapters 3, 5 and 6). Clues from recent studies suggest that certain phosphorylation and acetylation events of nuclear receptors can modulate protein–protein interactions of the receptor with the transcriptional machinery [4–7]. Similarly, sumolyation of the AR and GR may modulate receptor–protein interactions or receptor stability [8]. Studies with phosphorylation site-specific antibodies have revealed a role for site-specific phosphorylation and the intracellular location of the GR [9]. The continuing challenge is to determine the functional and/or structural significance of individual modifications and to determine if such modifications may actually act co-operatively.

"Action is transitory, — a step, a blow..." **William Wordsworth (*The Borders*)**

An important question when considering the mechanism of action of nuclear receptors is, are these proteins always nuclear? In the absence of a hormone, there is good evidence to suggest that a significant proportion of the receptors for androgens, glucocorticoids and progesterone are found in the cytoplasm, bound to a molecular chaperone complex (see Chapter 4). Molecular chaperones play important roles in steroid receptor signalling by maintaining the receptor in a conformation that can bind a hormone, in trafficking of the receptor complex from the cytoplasm to the nucleus and in the possible recycling of the receptor protein to switch off the response (see Chapters 4 and 6). In addition to the intracellular location of the inactive steroid receptors, recent studies have demonstrated that rapid responses to progesterone and oestradiol involve the extranuclear action of the classical receptor proteins (i.e. PR and ER) (see Chapter 8). The evidence highlights the action of these receptors at the plasma membrane, resulting in the activation of the tyrosine kinase c-src and the subsequent downstream mitogen-activated protein kinase signalling pathway.

The principal function of this family of proteins remains the direct regulation of gene transcription, which thus necessitates the nuclear location of the protein. Transcriptional regulation is a complex process that is mediated at a number of levels in eukaryotic cells. Not only is the transcription process itself subject to direct regulation, but the structure of the genetic material is modulated so as to permit or impede transcription via the packaging of chromosomal DNA with histone proteins as nucleosomes (see Chapter 6). Increasingly, attention is now being focused on investigating how these ligand-activated transcription factors regulate gene expression. A major breakthrough has been the identification of a diverse range of proteins that interact with nuclear receptors in either the presence (see Chapter 6) or absence (see Chapter 7) of specific ligands. Of particular interest has been the realization that some of these proteins, either alone or as part of multi-subunit complexes, harbour enzymic activities. The most notable enzymes associated with nuclear receptors include histone acetyltransferases/deacetylases, methyltransferases, ATP-dependent helicases and kinases (see Chapter 6).

It appears that, broadly speaking, nuclear receptors activate gene expression in a two-step process involving (i) opening of the chromatin structure, through histone modification and/or ATP-dependent remodelling of nucleosomes, and (ii) direct recruitment of the transcriptional machinery through interactions with key components including 'mediator' subunits, Med150 and Med220, and the basal transcription factors TBP (TATA-box-binding protein), transcription factor IIB (TFIIB), TFIIF and TFIIH (see Chapter 6). In contrast, the negative regulation of gene expression by nuclear receptors appears to involve (i) altered receptor–DNA interactions on so-called 'negative response elements', (ii) interactions with co-repressor complexes that lead to a closed chromatin structure, and (iii) protein–protein interactions that disrupt assembly of the transcriptional machinery or the action of other transcription factors (see Chapter 7). Techniques such as fluorescence recovery after photobleaching (FRAP) [10] and chromatin immunoprecipitation (ChIP) [11–13] have revealed the dynamic nature of receptor–DNA interactions and the cyclic recruitment of different co-activator complexes to natural target genes. The unravelling of the molecular details of these processes is very much an area of intense research in many laboratories.

"Accidents will occur in the best-regulated families" Charles Dickens (*David Copperfield*)

Given the important role that nuclear receptors play in development and regulatory processes in the adult animal, it is not surprising that disruption of these signalling pathways can have severe consequences for health. Mutations in the AR (see Chapter 9), GR (see Chapter 10), TR (see Chapter 12) and vitamin D3 receptor (reviewed in [14]) result in hormone resistance syndromes causing (i) in male differentiation and fertility, (ii) disruption of glucocorticoid homoeostasis, (iii) growth and developmental defects, and (iv) skeletal defects respectively. Resistance to androgens and vitamin D3 result from loss-of-function mutations clustering in the DBD and LBD, while general resistance to thyroid hormones results from mutations within the TRβ gene and involve the dominant negative action of the mutant receptor protein; the mutations having been mapped to discrete regions within the LBD of the three receptors. The identification of ligands for orphan receptors and the previously unsuspected signalling pathways has also opened up the possibility of genetic alterations in nuclear receptors in the underling pathologies of metabolic disorders such as diabetes and obesity (see Chapter 12).

Polymorphisms and mutations in the nuclear receptor genes have also been correlated with or been implicated in the risk of a wide range of conditions, including certain cancers, most notably breast (ER) and prostate cancer (AR), neurodegenerative disorders [ER, AR and NUR-related factor 1 (NURR1) /NGFI-B], cardiovascular disease (ER and TR) and a range of psychological conditions (ER and NURR1/NGFI-B) (see Chapters 9–12).

Considerable insight into the physiological roles of members of the nuclear receptor superfamily has also been gained from transgenic animal studies, including both 'knock-out' and 'knock-in' strategies (see Chapters 10–12). Such approaches have also be used successfully to investigate the role of co-activator and co-repressor proteins: for example, disruption of the gene for steroid receptor co-activator 1 (SRC-1) resulted in mice with symptoms of hormone-resistance syndromes (see Chapters 7 and 12). A better understanding of the mechanism of action of members of the nuclear receptor superfamily and the role that genetic alterations play in different diseases will ultimately yield important insights into the disease process and the development of new or improved therapeutic strategies that will target specific receptors in a tissue specific manner.

Conclusions and future perspectives

The last 20 years have seen tremendous progress in our understanding of the structure–function relationships of members of the nuclear receptor superfamily. This has included the isolation of receptor cDNAs, the solving at atomic resolution of the structure of isolated DBDs and LBDs, the identification of binding partners and a clearer appreciation of the role that mutations in nuclear receptors play in a wide range of pathological states. The future should be just as challenging and exciting, and we may reasonably expect progress in a number of areas.

- The three-dimensional structure of a full-length member of the nuclear receptor superfamily, probably in a complex with a DNA response element and/or peptides derived from co-regulatory proteins.
- A better understanding of the role of different post-translational modifications of nuclear receptors and co-regulatory proteins, and the enzymes responsible, in nuclear receptor signalling.
- Identification of gene targets and regulatory networks for individual nuclear receptors.
- A clearer understanding of the integration of different signalling pathways in cells, tissues and the whole organism in both health and disease states.

As befits such a large family of proteins that is involved in development, metabolic regulation and reproduction, there is intensive research into all aspects of nuclear receptors. In the future, it is assured that there will be further dramatic developments as researchers strive to understand the biology and physiology of nuclear receptors, the structural and biochemical basis for receptor action, and the role that genetic alterations in receptor signalling play in disease. The chapters that follow explore these and related aspects of nuclear receptor signalling in more detail, giving a good summary of our current knowledge and future directions in this ever-growing area of research.

Summary

- *Nuclear receptors represent a diverse group of transcription-regulating proteins that share a number of key structural and mechanistic properties that define this family of proteins.*
- *They contain an LBD consisting of a three-layer α-helix sandwich and a DBD consisting of a compact globular fold containing two perpendicular α-helices.*
- *They bind specific DNA response elements as homodimers (i.e. classical steroid receptors), heterodimers with RXR (i.e. steroid and non-steroid ligand receptors and orphan nuclear receptors) or monomers (i.e. certain orphan nuclear receptors).*
- *They activate transcription via (i) recruitment of co-activator complexes and (ii) direct interactions with the transcriptional machinery via basal transcription factors.*
- *They repress transcription by recruitment of co-repressor complexes, i.e. silencing mediator for RAR and TR (SMRT)- and nuclear receptor co-repressor (NcoR)-associated complexes.*
- *Receptor function can be regulated by molecular chaperone complexes and post-translational modifications, including phosphorylation, acetylation and sumoylation.*
- *Genetic alterations in nuclear receptors underpin disruption of reproductive function and metabolic homoeostasis, as well as neuro-degenerative disorders and certain cancers.*

Research in the author's laboratory is funded by the Association for International Cancer Research, the Biotechnology and Biological Sciences Research Council and the Medical Research Council.

References

1. Freeman, E.R., Bloom, D.A. & McGuire, E.J. (2001) A brief history of testosterone. *J. Urol.* **165**, 371–373
2. Jensen, E.V. & Jordan, V.C. (2003) The estrogen receptor: a model for molecular medicine. *Clin. Cancer Res.* **9**, 1980–1989
3. Evans, R.M. (1988) The steroid and thyroid hormone receptor superfamily. *Science* **240**, 889–895
4. Tremblay, A., Tremblay, G.B., Labrie, F. & Giguere, V. (1999) Ligand-independent recruitment of SRC-1 to estrogen receptor β through phosphorylation of activation function AF1. *Mol. Cell* **3**, 513–519
5. Hammer, G.D., Krylova, I., Zhang, Y., Darimont, B.D., Simpson, K., Weigel, N.L. and Ingraham, H.A. (1999) Phosphorylation of the nuclear receptor SF-1 modulates cofactor recruitment: integration of hormone signalling in reproduction and stress. *Mol. Cell* **3**, 521–526

6. Dutertre, M. & Smith, C.L. (2003) Ligand-independent interactions of p160/steroid receptor coactivators and CREB-binding protein (CBP) with estrogen receptor-α: regulation by phosphorylation sites in the A/B region depends on other receptor domains. *Mol. Endocrinol.* **17**, 1296–1314
7. Fu, M., Rao, M., Wang, C., Sakamaki, T., Wang, J., Di Vizio, D., Zhang, X., Albanese, C., Balk, S., Chang, C. et al. (2003) Acetylation of androgen receptor enhances coactivator binding and promotes prostate cancer cell growth. *Mol. Cell. Biol.* **23**, 8563–8575
8. Poukka, H., Karonen, U., Jänne, O.A. & Palvimo, J.J. (2000) Covalent modification of the androgen receptor by small ubiquitin-like modifier 1 (SUMO-1). *Proc. Natl. Acad. Sci. U.S.A.* **97**, 14145–14150
9. Wang, Z., Frederick, J. & Garabedian, M.J. (2002) Deciphering the phosphorylation "code" of the glucocorticoid receptor *in vivo*. *J. Biol. Chem.* **277**, 26573–26580
10. Hager, G.L., Fletcher, T.M., Xao, N., Baumann, C.T., Mueller, W.G. & McNally, J.G. (2000) Dynamics of gene targeting and chromatin remodelling by nuclear receptors. *Biochem. Soc. Trans.* **28**, 405–410
11. Shang, Y., Hu, X., DiRenzo, J., Lazar, M.A. & Brown, M. (2000) Cofactor dynamics and sufficiency in estrogen receptor-regulated transcription. *Cell* **103**, 843–852
12. Li, X., Jiemin, W., Tsai, S.Y., Tsai, M.-J. & O'Malley, B.W. (2003) Progesterone and glucocorticoid receptors recruit distinct coactivator complexes and promote distinct patterns of local chromatin modification. *Mol. Cell. Biol.* **23**, 3763–3773
13. Metivier, R., Penot, G., Huber, M.R., Reid, G., Brand, H., Kos, M. & Gannon, F. (2003) Estrogen receptor-α directs ordered, cyclical, and combinatorial recruitment of cofactors on a natural target promoter. *Cell* **115**, 751–763
14. Malloy, P.J., Pike, J.W. & Feldman D. (1999) The vitamin D receptor and the syndrome of hereditary 1,25-dihydroxyvitamin D-resistant rickets. *Endocr. Rev.* **20**, 156–188
15. Saporita, A.J., Zhang, Q., Navai, N., Dincer, Z., Hahn, J., Cai, X. & Wang, Z. (2003) Identification and characterization of a ligand-regulated nuclear export signal in the androgen receptor. *J. Biol. Chem.* **278**, 41998–42005

2

The evolution of the nuclear receptor superfamily

Héctor Escriva, Stéphanie Bertrand and Vincent Laudet[1]

UMR 5161 du CNRS, Ecole Normale Supérieure de Lyon, 46 allée d'Italie, 69364 Lyon Cedex 07, France

Abstract

Nuclear receptors form a superfamily of ligand-activated transcription factors implicated in various physiological functions from development to homoeostasis. Nuclear receptors share a common evolutionary history revealed by their conserved structure and by their high degree of sequence conservation. Here we review the lastest advances on the evolution of nuclear receptors by addressing the following questions. What is known about the appearance and diversification of nuclear homone receptors? How did their different functional characteristics evolve? What can we infer from the analysis of complete genomes? In summary, the study of the evolution of nuclear receptors will be very important not only for understanding their functions *in vivo* but also for obtaining insights into the evolution of animal genomes as a whole.

Introduction

Nuclear receptors (NRs) form a superfamily of ligand-activated transcription factors, which regulate various physiological functions, from development or reproduction to homoeostasis and metabolism, in animals (metazoans) [1]. The superfamily contains not only receptors for known ligands but also a large

[1]*To whom correspondence should be addressed (e-mail Vincent.laudet@ens-lyon.fr).*

number of so-called orphan receptors for which ligands do not exist or have not been identified. It is not known whether all of these orphan receptors indeed have a ligand, if they activate transcription in a constitutive manner or if they have alternative transcriptional regulation mechanisms [2].

NRs share a common structural organization with a central, well-conserved DNA-binding domain (DBD; also termed the C domain), a variable N-terminal region (A/B domain), a non-conserved hinge (D domain) and a C-terminal, moderately conserved, ligand-binding domain (LBD, or E domain) [1] (Figure 1). The three-dimensional structure has been determined for several NRs, unliganded (apo) or liganded (holo), allowing much better understanding of the mechanisms involved in ligand-binding and transactivation functions. These crystal structures show that the E domain undergoes a major conformational change upon ligand binding, allowing the interaction with co-activators and the transactivation of target genes [3,4].

NRs bind as homodimers or heterodimers to the regulatory regions of target genes, usually to the sequence PuGGTCA (where Pu is a purine), called hormone-response element (or HRE). However, mutation, extension and duplication, as well as distinct relative orientations of repeats of this motif, generate response elements that are selective for a given receptor or class of receptors [1] (see also Chapter 5 in this volume).

Liganded NRs include receptors for hydrophobic molecules such as steroid hormones (oestrogens, glucocorticoids, progesterone, mineralocorticoids, androgens, vitamin D, ecdysone, oxysterols, bile acids), retinoic acids, thyroid hormones, fatty acids, leukotrienes and prostagladins. The structural diversity of the ligands contrasts with the conservation and mode of action of their receptors. This and the large number of orphan receptors have prompted much speculation on the origin and evolution of the superfamily. The high degree of conservation between all NRs at the structural level (primary and

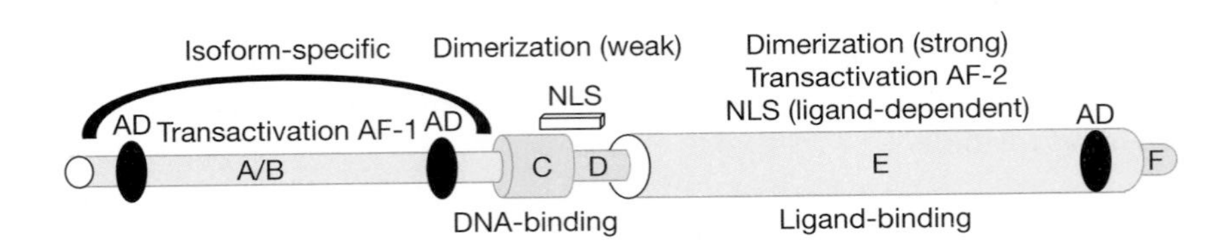

Figure 1. Structure of a bona fide NR, showing the various domains and their functions
Shown are the N-terminal domain (A/B domain), which is implicated in transactivation, the DBD (C domain), which is important for DNA-binding and dimerization functions, the variable hinge or D domain, the LBD (E domain), which is implicated in ligand binding, transcriptional regulation and dimerization, and the C-terminal domain or F domain, which is present in some but not all NRs. AD, activation domain; NLS, nuclear-localization signal; AF-2, activation function 2.

tertiary structure) suggests a common evolutionary origin from an ancestral NR whose origin is unknown.

Appearance and diversification of the NR superfamily

The first NR genes were cloned in the mid-1980s; the glucocorticoid receptor (GR) by Evans and colleagues [5] and the oestrogen receptor (ER) by the Chambon laboratory [6]. Since then, many other NRs have been isolated in most animal phyla, all of them showing a high degree of sequence conservation and the same conserved structure. This conservation, the important biological roles of NRs, and the facts that NR genes are good phylogenetic markers and are dispersed in animal genomes, make them an interesting subject for evolutionary studies [7].

Diversification of the NR superfamily

The phylogeny of the NR superfamily, using either the C or the E domain, separately or together, has been studied by several authors [8–11]. These studies suggest that the NRs evolved by duplication of a unique common ancestor that contained both the C and E domains. Thus the phylogenetic analysis of the superfamily resulted in a classification into six subfamilies of unequal size [12]. The phylogenetic relationships between subfamilies have been tested and confirmed by high bootstrap values in distance and parsimony analysis (Figure 2). This classification was used for an evolution-based nomenclature of the superfamily [13]. The six subfamilies are as follows.

1. A large subfamily containing thyroid hormone receptors (TRs), retinoic acid receptors (RARs), vitamin D receptor (VDR), ecdysone receptor (EcR) and peroxisome-proliferator-activated receptors (PPARs), as well as numerous orphan receptors [ROR (RAR-related orphan receptor), Rev-erb (reverse erbA), CAR (constitutive androstane receptor].
2. A subfamily containing retinoid X receptor (RXR), chicken ovalbumin upstream promoter-transcription factor (COUP-TF) and hepatocyte nuclear factor 4 (HNF4).
3. The steroid receptor subfamily with ER, GR, mineralocorticoid receptor (MR), progesterone receptor (PR) and androgen receptor (AR), as well as the ERR (oestrogen-receptor-related receptor) orphan receptors.
4. A subfamily containing the NGFI-B (nerve growth factor inducible factor I-B) group of orphan receptors,
5. The SF1 (steroidogenic factor 1) and *Drosophila* FTZ-F1 ('fushi tarazu' factor 1) subfamily.
6. A small subfamily containing only the GCNF1 (germ cell nuclear factor 1) receptor.

Since several subfamilies are present in early metazoans, this suggests that the superfamily underwent an explosive expansion during early metazoan

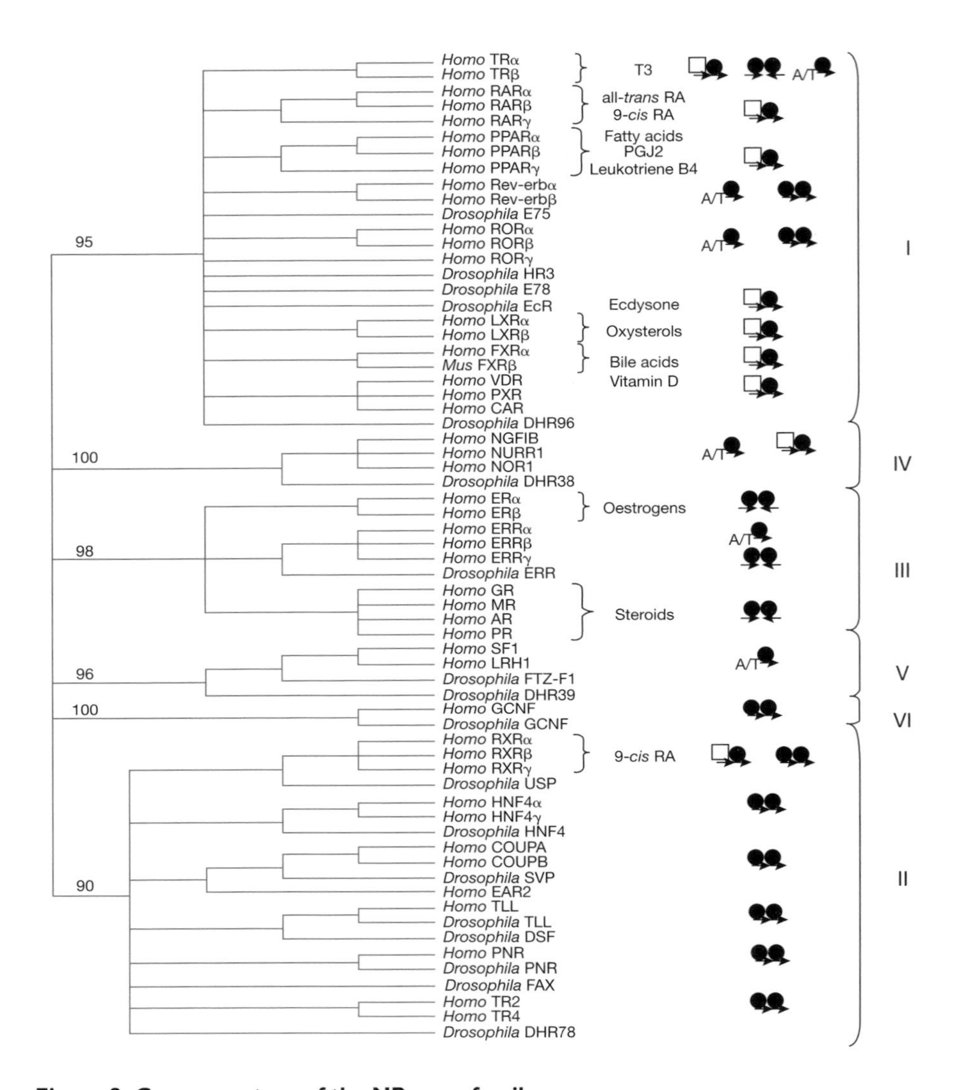

Figure 2. Consensus tree of the NR superfamily
Branches with bootstrap values under 90% were collapsed and presented as polytomies. The ligand-binding and dimerization abilities of NRs are schematized on the right. The arrowheads represent the PuGGTCA core motif, which can be a single unit, preceded by an A/T-rich sequence (A/T), a palindrome or a direct repeat. Black circles represent the receptors. The open squares represent RXR in the heterodimeric complexes. RA, retinoic acid; PGJ2, prostaglandin J_2; LXR, liver X receptor; HR3, hormone receptor 3; NOR1, neuron-derived orphan receptor 1; NURR1, NUR-related receptor 1; DHR, *Drosophila* hormone receptor; MR, mineralocorticoid receptor; AR, androgen receptor; LRH, liver receptor homologous protein; SVP, seven up; EAR, erbA-related receptor; TLL, tailless; PNR, photoreceptor-specific nuclear receptor.

evolution. Moreover, the diversification of the superfamily followed two waves of gene duplication: a first wave, before the deuterostome/protostome split, during the emergence of metazoans, led to the acquisition of the present six subfamilies and the various groups of receptors within each subfamily. The second wave occurred later on after the arthropod/vertebrate split, specifically in vertebrates, producing the paralogous groups within each subfamily (e.g. TRα and β, RARα, β and γ) [12,14] (Figure 3). This structure of the evolutionary tree is reminiscent of that observed for other gene families such as Hox or Ets [15,16]. This shows an early period of diversification during metazoan evolution (corresponding to the first wave of gene duplication previously mentioned for NRs) that is not correlated with any known genome duplications, which predates the protostome/deuterostome split, and a late period after the cephalochordate/vertebrate split (corresponding to the previously mentioned second wave of gene duplication of NRs) that is

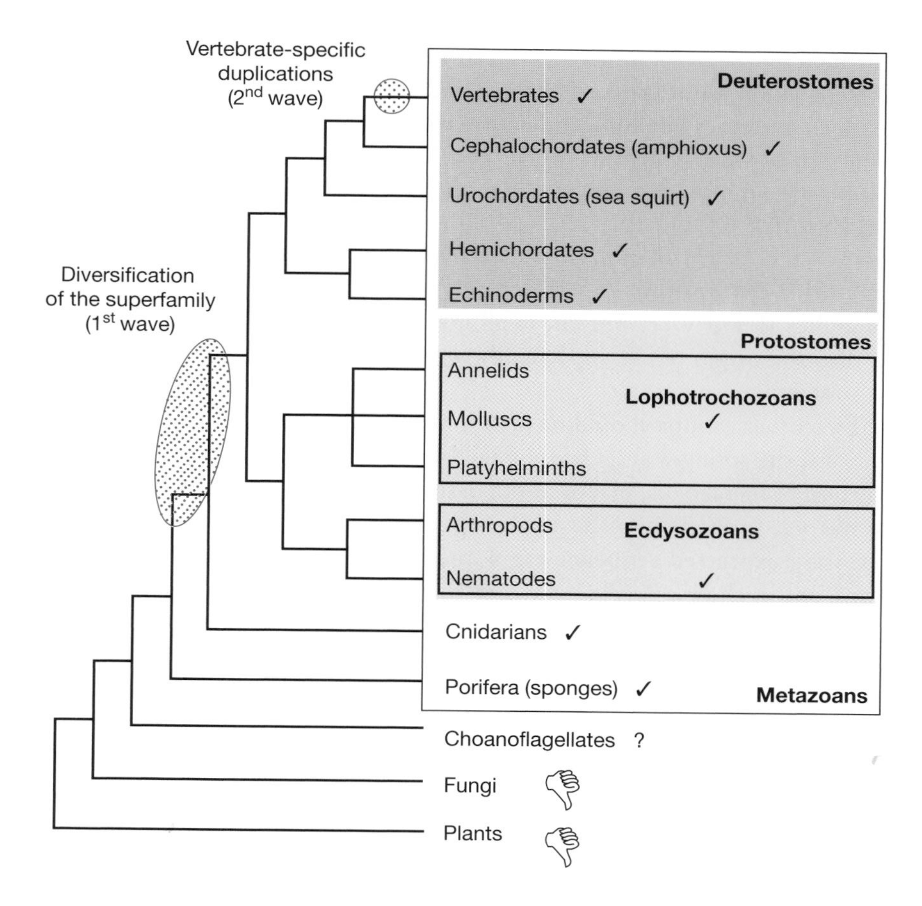

Figure 3. Schematic phylogeny of metazoans [diploblasts (cnidarians) and triploblasts (protostomes and deuterostomes)], plants, fungi and choanoflagellates
Clades where NRs have been isolated are labelled with a tick, clades where complete genome sequences exist without any NR gene are labelled with a 'thumb down' and clades where the presence of NRs is not yet known are labelled with ?. The two moments during the NR evolutionary history where big waves of gene duplication have been detected are also indicated.

correlated with the genome duplications, initially suggested by Ohno [17], that took place during vertebrate evolution.

Appearance of the NR superfamily

From an evolutionary point of view, the first questions to be addressed about the origin of the NR superfamily are when and how this superfamily appeared during evolution. In order to answer the first of these questions, we performed a systematic search for NRs by amplifying fragments of the most conserved region, the DBD or C region, in all the animal phyla (metazoa), but also in fungi, plants, algae and protozoa. Our analysis demonstrates that NRs are specific to metazoans since they were found in all the animal phyla tested from diploblastic animals (e.g. cnidarians) [14,18] to vertebrates, and not in plants, fungi, algae or protozoa (Figure 3). These results have subsequently been confirmed by the development of the complete genome-sequencing projects of different eukaryotes, since NRs have not been found in any of the non-metazoan genomes (i.e. unicellular eukaryotes, plants and fungi). Moreover, only members of subfamilies II and V (Figure 2) have been found in diploblastic animals [i.e. orthologues of the RXR (NR2B), TR2/4 (NR2C), TLL (NR2E), COUP (NR2F) and FTZ-F1 (NR5A) classes in diploblastic species such as *Hydra vulgaris*, *Montipora verrucosa*, *Anemonia sulcata*, *Tripedalia cystophora* and *Acropora millepora*] [14,18,19]. Thus only subfamilies II and V are present in all metazoans, suggesting that the first NR and thus the origin of the superfamily would probably belong to one of these two subfamilies.

Two other groups should be mentioned here due to their key evolutionary positions, the sponges and choanoflagellates. The choanoflagellates are the sister group of metazoans, placed at the base of metazoan diversification [20]. No one has yet looked for NRs in the choanoflagellate genome. However, an interesting expressed sequence tag sequencing programme is currently being carried out in choanoflagellates and in the near future we will probably know if NRs are present in these organisms. This will be an important issue for understanding the appearance of the superfamily (Figure 3).

Sponges are at the base of the metazoan tree and their evolutionary relationships are still debated [21]. We searched for an NR signature by genomic PCR in sponges without any clear positive result [14]. This led us to question the existence of NRs in sponges. Interestingly, an NR has been isolated very recently in the sponge *Suberites domuncula*, demonstrating that the superfamily already existed within the first metazoans [22]. This NR seems to be clearly a member of subfamily II, even if its precise relationships with the various groups (e.g. RXR, HNF4) of this subfamily is still unclear.

Concerning the question of how the first NR appeared during evolution, no clear signature reminiscent of NRs has been found in other protein families. However, low scores of sequence identity for the LBD and DBD regions of NRs with a peroxisomal membrane protein, Pex11p, and the LIM/GATA zinc-finger domain, respectively, have prompted some groups to suggest that the first NR was constructed by the fusion of two genes encoding these various proteins [23–25]. In the absence of structural data that can confirm the significance of these low similarity scores, it is hard to draw firm conclusions on this matter. The question of the origin of the first NR is thus still open.

Evolution of the NRs functional characteristics

From a schematic presentation of how an NR functions, it is easy to notice the different functional characteristics that are subject to evolutionary pressure. The main NR functional characteristics are the mode of DNA recognition (i.e. monomers, homodimers or heterodimers with RXR), the different HREs recognized (i.e. palindromes, direct repeats, inverted repeats), the different ligands bound by the NRs, or their orphan versus liganded status, and the co-repressor–co-activator interactions. In the following discussion we will examine these various functional characteristics of NRs from an evolutionary point of view.

Evolutionary studies of the interaction of NRs with co-repressors or co-activators have not yet been performed, and thus we do not know whether these co-regulators and these protein–protein interactions are evolutionarily conserved or not. The only NR-specific co-regulators known to date from non-vertebrate organisms are a co-repressor (silencing mediator of repressed transcription, or SMRTER) [26] and a co-activator (Taiman) [27] in *Drosophila melanogaster*. Both SMRTER and Taiman show a low degree of sequence conservation when compared with their vertebrate orthologues, nuclear receptor co-repressor (NCoR) and the p160/steroid receptor co-activator family (SRC) respectively. Moreover, in both cases the conservation is restricted to specific domains of the two proteins and not to the whole length. SMRTER shows 35% sequence conservation with the SNOR domain of NCoR and Taiman shows 50% sequence conservation with the basic helix–loop–helix domain of steroid receptor co-activator proteins. It is also well known that NRs are able to activate transcription in yeast despite the fact that there are no NR genes in its genome. However, if we look, using sequence comparison, for NR-specific co-regulators within the yeast genome, we do not obtain any clear orthologue. On the other hand, it has been demonstrated that GR and ER interact directly with the yeast proteins SWI/SNF and Spt6. These are proteins from the yeast Ada complex. The Ada complex activates transcription, either through the modification of chromatin structure via the histone acetyltranferase activity of its subunit GCN5 (a yeast transcriptional co-activator), or by interacting with the TATA-box-binding protein in the proximal promoter region. We observe that the yeast GCN5 protein shows 40% sequence identity with the human p300/CBP-associated factor [where CBP means cAMP-response-element-binding protein (CREB)-binding protein]. For SWI/SNF, the human orthologues (i.e. chromatin-remodelling factors BRG1 and BRM) show 30% sequence identity with the yeast protein. But the co-regulators p300/CBP-associated factor, BRG1 and BRM are not specific for NRs and are also involved in other gene-expression-control pathways. All these observations suggest that the first NR, as well as other transcription factors, recruited pre-existing transcriptional regulators in the common ancestor of all metazoans to regulate transcription, and later, the NR-specific co-regulators evolved specifically in metazoans.

What do we find if we correlate the other functional characteristics of NRs with their evolutionary history? Some of the characteristics of NRs, such as dimerization and DNA-binding abilities, correlate well with the phylogenetic position of a given NR within the tree (Figure 2). For example, steroid receptors, such as ER or GR, all bind as homodimers to palindromic elements. The closely related Rev-erb and ROR orphan receptors bind as monomers or dimers to the same response elements. Most if not all RXR-interacting receptors belong to subfamilies I (e.g. TR, RAR, PPAR and VDR) and IV (NGFI-B), suggesting that this characteristic is a shared derived character (i.e. synapomorphy) uniting these two subfamilies. This correlation between the phylogenetic position of a given NR and some functional characteristics suggests a common evolutionary history.

Strikingly, such a correlation is not found when the ligand-binding abilities of the receptors are compared with phylogeny (Figure 2). The phylogenetic position of a receptor is clearly not correlated with the chemical nature of its ligand or even with its liganded versus orphan status. For example, the evolutionarily closely related receptors of subfamily I (TRs, RARs, PPARs and VDRs) bind ligands originating from totally different biosynthetic pathways. Conversely, RAR (subfamily I) and RXR (subfamily II), which are not evolutionarily closely related, bind similar ligands (all-*trans* and 9-*cis* retinoic acid respectively). And finally, orphan receptors are not clustered in a specific subfamily, but spread through the whole phylogenetic tree of the NR superfamily. This situation could be explained by an independent gain of ligand-binding capacity several times during NR evolution, specifically in each branch of the tree [2] (Figure 4). This hypothesis implies that the ancestral NR was orphan and thus its activation was not produced by the ligand binding. It is now well accepted that signals other than the ligand itself can regulate NR activity [28]. For example, the epidermal growth factor receptor signalling activates ER and its target genes in an oestrogen-independent manner through a specific phosphorylation of serine 118 of ER [29]. Similarly, dopamine is also capable of activating PR in a ligand-independent manner [30]. The three-dimensional structure of the orphan receptor NUR-related receptor 1 (NURR1) provides an example of a LBD that does not have enough space for a ligand, even if its native conformation parallels that of the active form of liganded NRs. Moreover, NURR1 activity is dependent on its AF2-AD (activation function 2-activation domain) and its regulation is associated with a ligand-independent structural transition of its LBD, similar to the apo-to-holo transition of liganded receptors. This structural transition is probably regulated by a crosstalk with

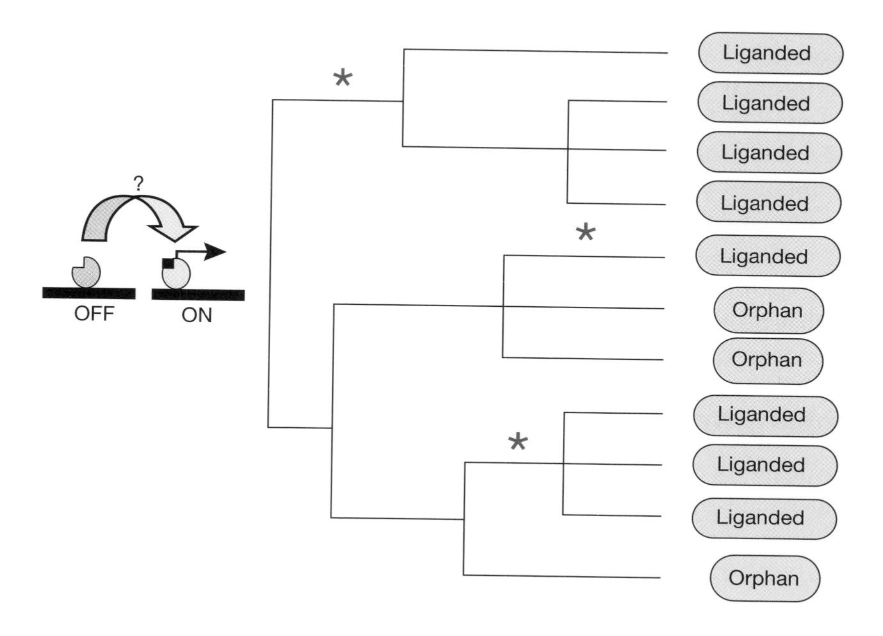

Figure 4. Schematic model for ligand-binding acquisition through evolution
The ancestral orphan receptor (with an unknown activation method) gained ligand-binding capacity (asterisks) and the new functionality was fixed in its descendents. It should be noted that for each NR ancestor, ligand-binding acquisition may have been gained at different moments of evolution, as shown by the location of the asterisks.

the mitogen-activated protein kinase signalling pathway [31]. This example strongly suggests that authentic orphan receptors, regulated by a ligand-independent conformational change, really exist. Other possibilities such as the interaction of the NR LBD with other proteins, resulting in an appropriate conformation for interaction with co-activators, cannot be excluded. For example, it has been shown that the bicoid-related homeobox protein Ptx1 regulates the orphan NR SF1 through an interaction that mimics the role of the ligand [32]. Finally, it should be noted here that orphan receptors usually contain a conserved LBD with an AF2-AD domain and a conserved three-dimensional structure [4], suggesting that an evolutionary pressure (other than ligand binding) exists for the conservation of such a structure.

It has recently been shown that some orphan NRs, such as HNF4 [33] or USP (ultraspiracle) [34,35], constitutively bind fatty acids. USP is in a constitutive inactive conformation and HNF4 is a constitutive activator. Taking these new data into account, it has been proposed that the original NR did bind a lipophilic compound as an integral component of the protein structure [36]. The constitutively bound fatty acid in that case could not be considered as a real ligand that is able to induce a switch on the activity of the receptor. So the question is: what is a ligand? An NR ligand is a small hydrophobic molecule that binds directly to the LBD and promotes a structural rearrangement allowing the AF2-AD to interact with transcriptional co-regulators that will connect the NR to the basic transcriptional machinery, resulting in the transcriptional stimulation/silencing of target genes. In this sense, the fatty acids found in USP or HNF4 should be considered as cofactors rather than ligands, similar to the haem group of haemoglobin. Moreover, if the ancestral NR contained a structural molecule (like USP and HNF4), this structural constraint could help to explain the high level of sequence and structure conservation between all NRs. From this, we can easily hypothesize how real ligand regulation evolved: from cofactors such as the fatty acid in the HNF4 LBD that provide a previous step in the evolutionary pathway, from ligand-independent to real ligand-regulated receptors.

Comparative genomics of NRs

The availability of complete genome sequences provides a unique opportunity to understand the various steps that have led to the present-day diversity of NRs. From these data, it is clear that fungi (*Saccharomyces cerevisiae* [37], *Schizosaccharmyces pombe* [38], *Encephalitozoon cuniculi* [39]), plants (*Arabidopsis thaliana* [40], *Oryza sativa* [41]) or protists (*Plasmodium falciparum* [42]) do not contain NRs, even if some short matches with high scores were reported with FASTA homology searches. This confirms that NRs are specific to metazoans.

In the last few years, complete genome sequences have been determined for different species of metazoans: *Caenorhabditis elegans* [43] (a nematode), *Drosophila* [44] and *Anopheles* [45] (insects), *Ciona intestinalis* [46] (urochordate), *Takifugu rubripes* [47] and *Tetraodon nigroviridis* [48] (fishes), mouse [49] and human [50] (mammals). It is now possible to study the whole set of NRs for a given organism in a systematic way. We can thus study the coding sequences, the gene structure, the chromosomic position and the promoter sequences for all the NRs of a single organism. Moreover, the availability of complete genome sequences allows the identification of all the members of the NR superfamily for each species without limitations imposed by temporal, spatial or quantitative aspects of mRNA expression. The whole set of NRs has been reported for some of these metazoans, such as *Drosophila*, *C. elegans* and human with some striking results.

The *Drosophila* genome contains a low number of NRs (21; including four previously unknown, ERR, GCNF, photoreceptor-specific nuclear receptor and one that lacks a mammalian orthologue, DSF). The situation is very different in the genome of the nematode worm *C. elegans*, the first metazoan whose genome was completely sequenced [43]. Indeed, about 270 NR sequences have been found in this genome [51]. If clear homologues with mammalian or *Drosophila* sequences were found for a minority of these receptors, most of them share an evolutionary origin and harbour no peculiar similarity with NRs from other organisms (Robinson-Rechavi, M., Maina, C.V., Gissendanner, C.R., Laudet, V.

and Sluder, A., unpublished work). This suggests that these numerous receptors evolved through an extensive series of gene duplications that arose specifically in the nematode lineage [52]. The reason for this striking proliferation is still a matter of speculation. In contrast to the nematode, and in accordance with the data available in *Drosophila*, the human genome appears to contain 48 NRs [53]. Moreover, the study of the human genome allowed us to discover a previously unknown FXRβ (farnesoid receptor β)gene that shows a high degree of sequence conservation with FXRα as well as conservation of the exon–intron structure, but some in-frame stop codons. Expressed Sequence Tags corresponding to this FXRβ were also found, allowing us to hypothesize that this FXRβ gene may have lost its function in humans. Indeed, later on, an FXRβ orthologue has been found as a functional gene in the mouse and other mammals [54]. Thus the results obtained with FXRα-knockout mice should be revisited, since the signalling pathway in which the FXRs are implicated is not completely abrogated in these animals. The mouse thus contains 49 functional NR genes.

Differences between NRs from insects, the nematode and mammals are extremely interesting and can provide important clues to the understanding of the evolution of the superfamily and some specific functions such as the ligand-binding capacity. For example, NRs such as RAR, TR or PPAR have only been found in deuterostomes and are not present in *Drosophila* or *C. elegans*. But is this absence due to a specific loss of such receptors in these two extremely divergent animals, or did RAR, TR and PPAR appear specifically in the deuterostome lineage?

However, due to the evolutionary distance between protostomes and humans, the study of protostomian NRs cannot be easily extrapolated to humans (e.g. role in development, implication in diseases such as cancer or diabetes). Thus genomes from other deuterostomes that are evolutionarily closer to humans will be more informative for the comparative study of NR function in vertebrates. Thus sequenced genomes of vertebrates such as human, mouse and the fishes *Fugu* and *Tetraodon* will allow us to compare the NR enhancer regions and discover important motifs for the conserved functions of NRs in vertebrates.

Another recently reported deuterostome genome sequence is that of the sea squirt (*C. intestinalis*). Sea squirts or ascidians are deuterostome metazoans distantly related to humans. Up to now, they represent the closest sequenced invertebrate genome to the vertebrates. A complete set of 17 NRs (plus one pseudogene) has been reported in *C. intestinalis*, containing a representative of most of the human NR subfamilies that are not present in *Drosophila*, such as RAR, PPAR or TR, suggesting that the ancestral functions controlled by these receptors evolved together with the deuterostome evolutionary lineage. But the absence of an orthologue of the steroid receptors (AR, MR, GR, PR and ER) raises the question of the evolution of the steroid signalling pathways in vertebrates.

Conclusions and perspectives

Great amounts of data concerning expression, function, interactions and evolution have been accumulated in the last two decades for many NRs in mammals. However, the more we know, the more questions that arise about the different functional properties of NRs. One of the most studied questions that will certainly be strongly developed in the next years concerns the mode of action of orphan receptors. Do they have a natural ligand? If not, how are they activated? How is their activity modulated? One important way to understand the function of orphan receptors will probably be the comparative study of their activity and their three-dimensional structures with liganded NRs in their apo and holo forms.

The new tools available in this 'genomics' era will certainly help to clarify the functions of NRs *in vivo*. For example, the bioinformatic comparison of regulatory regions of orthologues and paralogues from different species (i.e. the so-called phylogenetic footprinting) is likely to identify new enhancers. Expression studies of the whole set of NRs and their co-regulators for a specific organism at different physiological status or following different pharmacological treatments, by using DNA chips, will show how and which NRs and co-regulators are up- or down-regulated, allowing new hypotheses to be tested about their interactions.

From an evolutionary point of view, phylogenetic studies have used specific genes to understand either the evolution of a gene family or to infer the evolution of species. However, an area of much current interest is the mechanism of metazoan genome evolution. In this context, the high degree of sequence conservation between NRs (i.e. it is easy to clone NRs from different metazoans by PCR), their dispersed position in animal genomes (thus representing an unbiased sample of sequences) and their duplication history, make an excellent tool for the study of the evolution of animal genomes as a whole [7].

A new field in which NRs will certainly play an important role in the coming years is the field of comparative genomics. The use of evolutionary related organisms, placed at key positions of the phylogenetic tree, as we have seen with the sea squirt, will allow us to understand the appearance and evolution of important hormonal regulatory pathways (e.g. the steroids). In the future, these genomic studies should be extended to other species that could be even more informative. Thus to understand the evolution of steroid regulatory pathways, the characterization of NRs of the two main evolutionary lineages that diverged between sea squirts (with no steroid receptors) and craniates (with the four classical steroid receptors; AR, MR, GR and PR) will be very helpful. The first is the closest invertebrate to vertebrates, the amphioxus, which is much less divergent compared with vertebrates than ascidians. The amphioxus whole NR data set will give us the answer about the appearance of steroid receptors in the vertebrate lineage. And the second key organism is the lamprey (in which two steroid receptors have been isolated [55, 56]). The functional characterization of steroid receptors within the deuterosmes [none in ascidian, not known in amphioxus, two in lampreys and four (AR, GR, PR and MR) in vertebrates] will shed light on the evolution of the ligand-binding specificity of the four craniate steroid receptors and the evolution of their physiological roles.

Concerning the other main triploblastic lineage, the protostomes (Figure 3), the study of NRs from a lophotrocozoan and comparison with NRs from the ecdysozoans (i.e. *Anopheles*, *Drosophila*), will provide further insight into how NRs evolved in protostomes. Do both lineages share the same set of NRs, meaning that probably they also share similar regulatory pathways? One example is the only protosmian receptor with a known ligand (20-hydroxyecdysone for EcR). The question is open about the evolution of ecdysone-controlled metamorphosis (ecdysis) in protostomes. Even though a peak of ecdysone precedes metamorphosis in the lophotrocozoan *Schistosoma mansoni* [57], as occurs in ecdysozoa, we failed to isolate an EcR in this trematode [14]. So, apparently EcR is specific to ecdysozoans, and we could hypothesize that even if the ecdysone-controlled metamorphosis predates the evolution of EcR, EcR appeared and evolved specifically in ecdysozoans, subsequently acquiring the capacity to bind ecdysone and to control metamorphosis [58]. The only way to answer this question directly will be to demonstrate the absence of EcR in lophotrocozoans by a whole-genome-sequence approach.

Summary

- *NRs form a superfamily of transcription factors present in all the metazoan phyla that share a common evolutionary origin.*
- *The superfamily of NRs evolved following two waves of gene duplication. The first, very early during metazoan evolution, resulted in the diversification into six subfamilies and the second, specifically in vertebrates, produced the different paralogues within each subfamily.*
- *The ligand-binding function of NRs was probably acquired during metazoan evolution. On the contrary, DNA-binding and dimerization functions co-evolved with NRs.*
- *Complete genome sequences of C. elegans, D. melanogaster, C. intestinalis, Mus musculus and Homo sapiens have permitted the identification of approx. 270, 21, 18, 49 and 48 NRs respectively. The number and type of NRs in each genome depends on the species' phylogenetic position (deuterostome versus protostome) and particular evolutionary history.*

References

1. Laudet, V. & Gronemeyer, H. (2002) *The Nuclear Receptors, Factsbook*, Academic Press, London
2. Escriva, H., Delaunay, F. & Laudet, V. (2000) Ligand binding and nuclear receptor evolution. *Bioessays* **22**, 717–727
3. Moras, D. & Gronemeyer, H. (1998) The nuclear receptor ligand-binding domain: structure and function. *Curr. Opin. Cell Biol.* **10**, 384–391
4. Wurtz, J.M., Bourguet, W. Renaud, J.P., Vivat, V., Chambon, P., Moras, D. & Gronemeyer, H. (1996) A canonical structure for the ligand-binding domain of nuclear receptors. *Nat. Struct. Biol.* **3**, 87–94
5. Hollenberg, S.M. Weinberger, C., Ong, E.S., Cerelli, G., Oro, A., Lebo, R., Thompson, E.B., Rosenfeld, M.G. & Evans, R.M. (1985) Primary structure and expression of a functional human glucocorticoid receptor cDNA. Nature (London) **318**, 635–641
6. Green, S., Walter, P., Kumar, V., Krust, A., Bornert, J.M., Argos, P. & Chambon, P. (1986) Human oestrogen receptor cDNA: sequence, expression and homology to v-erb-A. *Nature (London)* **320**, 134–139
7. Escriva, H., Laudet, V. & Robinson-Rechavi, M. (2003) Nuclear receptors are markers of animal genome evolution. *J. Struct. Funct. Genomics* **3**, 177–184
8. Laudet, V., Hanni, C., Coll, J., Catzeflis, F. & Stehelin, D. (1992) Evolution of the nuclear receptor gene superfamily. *EMBO J.* **11**, 1003–1013
9. Amero, S.A., Kretsinger, R.H., Moncrief, N.D., Yamamoto K.R. & Pearson, W.R. (1992) The origin of nuclear receptor proteins: a single precursor distinct from other transcription factors. *Mol. Endocrinol.* **6**, 3–7
10. Detera-Wadleigh, S.D. & Fanning, T.G. (1994) Phylogeny of the steroid receptor superfamily. *Mol. Phylogenet. Evol.* **3**, 192–205
11. Thornton, J.W. & DeSalle, R. (2000) A new method to localize and test the significance of incongruence: detecting domain shuffling in the nuclear receptor superfamily. *Syst. Biol.* **49**, 183–201
12. Laudet, V. (1997) Evolution of the nuclear receptor superfamily: early diversification from an ancestral orphan receptor. *J. Mol. Endocrinol.* **19**, 207–226
13. Nuclear Receptors Nomenclature Committee (1999) A unified nomenclature system for the nuclear receptor superfamily. *Cell* **97**, 161–163
14. Escriva, H., Safi, R., Hanni, C., Langlois, M.C., Saumitou-Laprade, P., Stehelin, D., Capron, A., Pierce, R. & Laudet, V. (1997) Ligand binding was acquired during evolution of nuclear receptors. *Proc. Natl. Acad. Sci. U.S.A.* **94**, 6803–6808
15. Holland, P.W. & Garcia-Fernandez, J. (1996) Hox genes and chordate evolution. *Dev. Biol.* **173**, 382–395
16. Laudet, V., Hanni, C., Stehelin, D. & Duterque-Coquillaud, M. (1999) Molecular phylogeny of the ETS gene family. *Oncogene* **18**, 1351–1359
17. Ohno, S. (1970) *Evolution by Gene Duplication*, Springer-Verlag, Heidelberg
18. Grasso, L.C., Hayward, D.C., Trueman, J.W., Hardie, K.M., Janssens P.A. & Ball, E.E. (2001) The evolution of nuclear receptors: evidence from the coral *Acropora*. *Mol. Phylogenet. Evol.* **21**, 93–102
19. Kostrouch, Z., Kostrouchova, M., Love, W., Jannini, E., Piatigorsky, J. & Rall, J.E. (1998) Retinoic acid X receptor in the diploblast, *Tripedalia cystophora*. *Proc. Natl. Acad. Sci. U.S.A.* **95**, 13442–13447
20. Snell, E.A., Furlong, R.F. & Holland, P.W. (2001) Hsp70 sequences indicate that choanoflagellates are closely related to animals. *Curr. Biol.* **11**, 967–970
21. Cavalier-Smith, T., Allsopp, E.P., Chao, E.E., Boury-Esnault, N. & Vacelet, J. (1996) Sponge phylogeny, animal monophyly, and the origin of the nervous system: 18S rRNA evidence. *Can. J. Zool.* **74**, 2031–2045
22. Wiens, M., Batel, R., Korzhev, M. & Muller, W.E. (2003) Retinoid X receptor and retinoic acid response in the marine sponge *Suberites domuncula*. *J. Exp. Biol.* **206**, 3261–3271
23. O'Malley, B.W. (1989) Did eucaryotic steroid receptors evolve from intracrine gene regulators? *Endocrinology* **125**, 1119–1120

24. Barnett, P., Tabak, H.F. & Hettema, E.H. (2000) Nuclear receptors arose from pre-existing protein modules during evolution. *Trends Biochem. Sci.* **25**, 227–228
25. Clarke, N.D. & Berg, J.M. (1998) Zinc fingers in *Caenorhabditis elegans*: finding families and probing pathways. *Science* **282**, 2018–2022
26. Tsai, C.C., Kao, H.Y., Yao, T.P., McKeown, M. & Evans, R.M. (1999) SMRTER, a *Drosophila* nuclear receptor coregulator, reveals that EcR-mediated repression is critical for development. *Mol. Cell* **4**, 175–186
27. Bai, J., Uehara, Y. & Montell, D.J. (2000) Regulation of invasive cell behavior by taiman, a *Drosophila* protein related to AIB1, a steroid receptor coactivator amplified in breast cancer. *Cell* **103**, 1047–1058
28. Cenni, B. & Picard, D. (1999) Ligand-independent activation of steroid receptors: new roles for old players. *Trends Endocrinol. Metab.* **10**, 41–46
29. Curtis, S.W., Washburn, T., Sewall, C., DiAugustine, R., Lindzey, J., Couse, J.F. & Korach, K.S. (1996) Physiological coupling of growth factor and steroid receptor signaling pathways: estrogen receptor knockout mice lack estrogen-like response to epidermal growth factor. *Proc. Natl. Acad. Sci. U.S.A.* **93**, 12626–12630
30. Mani, S.K., Allen, J.M., Lydon, J.P., Mulac-Jericevic, B., Blaustein, J.D., DeMayo, F.J., Conneely, O. & O'Malley, B.W. (1996) Dopamine requires the unoccupied progesterone receptor to induce sexual behavior in mice. *Mol. Endocrinol.* **10**, 1728–1737
31. Wang, Z., Benoit, G., Liu, J., Prasad, S., Aarnisalo, P., Liu, X., Xu, H., Walker, N.P. & Perlmann, T. (2003) Structure and function of Nurr1 identifies a class of ligand-independent nuclear receptors. *Nature (London)* **423**, 555–560
32. Tremblay, J.J., Marcil, A., Gauthier, Y. & Drouin, J. (1999) Ptx1 regulates SF-1 activity by an interaction that mimics the role of the ligand-binding domain. *EMBO J.* **18**, 3431–3441
33. Wisely, G.B., Miller, A.B., Davis, R.G., Thornquest, Jr, A.D., Johnson, R., Spitzer, T., Sefler, A., Shearer, B., Moore, J.T., Willson, T.M. et al. (2002) Hepatocyte nuclear factor 4 is a transcription factor that constitutively binds fatty acids. *Structure (Cambridge)* **10**, 1225–1234
34. Billas, I.M., Moulinier, L., Rochel, N. & Moras, D. (2001) Crystal structure of the ligand-binding domain of the ultraspiracle protein USP, the ortholog of retinoid X receptors in insects. *J. Biol. Chem.* **276**, 7465–7474
35. Clayton, G.M., Peak-Chew, S.Y, Evans, R.M. & Schwabe, J.W. (2001) The structure of the ultraspiracle ligand-binding domain reveals a nuclear receptor locked in an inactive conformation. *Proc. Natl. Acad. Sci. U.S.A.* **98**, 1549–1554
36. Sladek, F.M. (2002) Desperately seeking...something. *Mol. Cell* **10**, 219–221
37. Goffeau, A., Barrell, B.G., Bussey, H., Davis, R.W., Dujon, B., Feldmann, H., Galibert, F., Hoheisel, J.D., Jacq, C., Johnston, M. et al. (1996) Life with 6000 genes. *Science* **274**, 546, 563–567
38. Wood, V., Gwilliam, R., Rajandream, M.A., Lyne, M., Lyne, R., Stewart, A., Sgouros, J., Peat, N., Hayles, J., Baker, S. et al. (2002) The genome sequence of *Schizosaccharomyces pombe. Nature (London)* **415**, 871–880
39. Katinka, M.D., Duprat, S., Cornillot, E., Metenier, G., Thomarat, F., Prensier, G., Barbe, V., Peyretaillade, E., Brottier, P., Wincker, P. et al. (2001) Genome sequence and gene compaction of the eukaryote parasite *Encephalitozoon cuniculi. Nature (London)* **414**, 450–453
40. Arabidopsis Genome Initiative (2000) Analysis of the genome sequence of the flowering plant *Arabidopsis thaliana. Nature (London)* **408**, 796–815
41. Yu, J., Hu, S., Wang, J., Wong, G.K., Li, S., Liu, B., Deng, Y., Dai, L., Zhou, Y., Zhang, X. et al. (2002) A draft sequence of the rice genome (*Oryza sativa* L. ssp. *indica*). *Science* **296**, 79–92
42. Gardner, M.J., Hall, N., Fung, E., White, O., Berriman, M., Hyman, R.W., Carlton, J.M., Pain, A., Nelson, K.E., Bowman, S. et al. (2002) Genome sequence of the human malaria parasite *Plasmodium falciparum. Nature (London)* **419**, 498–511
43. The *C. elegans* Sequencing Consortium (1998) Genome sequence of the nematode *C. elegans*: a platform for investigating biology. *Science* **282**, 2012–2018

44. Adams, M.D., Celniker, S.E., Holt, R.A., Evans, C.A., Gocayne, J.D., Amanatides, P.G., Scherer, S.E., Li, P.W., Hoskins, R.A., Galle, R.F. et al. (2000) The genome sequence of *Drosophila melanogaster*. *Science* **287**, 2185–2195
45. Holt, R., Subramanian, G., Halpern, A., Sutton, G., Charlab, R., Nusskern, D.R., Wincke, P., Clark, A.G., Ribeiro, J.M., Wides, R. et al. (2002) The genome sequence of the malaria mosquito *Anopheles gambiae*. *Science* **298**, 129–149
46. Dehal, P., Satou, Y., Campbell, R.K., Chapman, J., Degnan, B., De Tomaso, A., Davidson, B., Di Gregorio, A., Gelpke, M., Goodstein, D.M. et al. (2002) The draft genome of *Ciona intestinalis*: insights into chordate and vertebrate origins. *Science* **298**, 2157–2167
47. Aparicio, S., Chapman, J., Stupka, E., Putnam, N., Chia, J.M., Dehal, P., Christoffels, A., Rash, S., Hoon, S., Smit, A. et al. (2002) Whole-genome shotgun assembly and analysis of the genome of *Fugu rubripes*. *Science* **297**, 1301–1310
48. Roest-Crollius, H., Jaillon, O., Bernot, A., Dasilva, C., Bouneau, L., Fischer, C., Fizames, C., Wincker, P., Brottier, P., Quetier, F. et al. (2000) Estimate of human gene number provided by genome-wide analysis using *Tetraodon nigroviridis* DNA sequence. *Nat. Genet.* **25**, 235–238
49. Waterston, R., Lindblad-Toh, K., Birney, E., Rogers, J., Abril, J.F., Agarwal, P., Agarwala, R., Ainscough, R., Alexandersson, M., An, P. et al. (2002) Initial sequencing and comparative analysis of the mouse genome. *Nature (London)* **420**, 520–562
50. Lander, E.S., Linton, L.M., Birren, B., Nusbaum, C., Zody, M.C., Baldwin, J., Devon, K., Dewar, K., Doyle, M., FitzHugh, W. et al. (2001) Initial sequencing and analysis of the human genome. *Nature (London)* **409**, 860–921
51. Sluder, A.E., Mathews, S.W., Hough, D., Yin V.P. & Maina, C.V. (1999) The nuclear receptor superfamily has undergone extensive proliferation and diversification in nematodes. *Genome Res.* **9**, 103–120
52. Sluder, A.E. & Maina, C.V. (2001) Nuclear receptors in nematodes: themes and variations. *Trends Genet.* **17**, 206–213
53. Robinson-Rechavi, M., Carpentier, A.S., Duffraisse, M. & Laudet, V. (2001) How many nuclear hormone receptors are there in the human genome? *Trends Genet.* **17**, 554–556
54. Otte, K., Kranz, H., Kober, I., Thompson, P., Hoefer, M., Haubold, B., Remmel, B., Voss, H., Kaiser, C., Albers, M. et al. (2003) Identification of farnesoid X receptor beta as a novel mammalian nuclear receptor sensing lanosterol. *Mol. Cell. Biol.* **23**, 864–872
55. Thornton, J.W. (2001) Evolution of vertebrate steroid receptors from an ancestral estrogen receptor by ligand exploitation and serial genome expansions. *Proc. Natl. Acad. Sci. U.S.A.* **98**, 5671–5676
56. Thornton, J.W., Need, E. & Crews, D. (2003) Resurrecting the ancestral steroid receptor: ancient origin of estrogen signaling. *Science* **301**, 1714–1717
57. Nirde, P., Torpier, G., De Reggi, M.L. & Capron, A. (1983) Ecdysone and 20 hydroxyecdysone: new hormones for the human parasite *Schistosoma mansoni*. *FEBS Lett.* **151**, 223–227
58. de Mendonca, R.L., Escriva, H., Vanacker, J.M., Bouton, D., Delannoy, S., Pierce, R. & Laudet, V. (1999) Nuclear hormone receptors and evolution. *Am. Zool.* **39**, 704–713

3

Overview of the structural basis for transcription regulation by nuclear hormone receptors

Raj Kumar, Betty H. Johnson, and E. Brad Thompson[1]

Department of Human Biological Chemistry & Genetics, University of Texas Medical Branch, 301 University Blvd., 5.104 Medical Research Bldg., Galveston, TX 77555, U.S.A.

Abstract

The mechanism of action of the nuclear hormone receptors (NHRs) as gene-regulatory molecules has become a major focus of current biological interest. NHRs belong to the superfamily of ligand-activated transcription factors, which are involved in the regulation of homoeostasis, reproduction, development and differentiation. To fully understand their functions, it is important to know the functional three-dimensional structure of these proteins. Molecular cloning and structure–function analyses have revealed that NHRs commonly have three functional regions: the N-terminal, DNA-binding and ligand-binding domains. Structures of some of these domains expressed independently have been solved. However, to date the three-dimensional structure remains unknown for full-length and even for any two domains together of any NHR family member. The available structures nevertheless begin to give clues of how site-specific DNA binding takes place, and how ligand binding alters the ligand-binding domain, consequently affecting potential interactions of the NHRs with co-

[1]*To whom correspondence should be addressed (e-mail bthompso@utmb.edu).*

activators/co-repressors and other components of basal transcriptional machinery. However, precisely how signals from a ligand through its NHR are passed to specific genes is still unknown. Herein, we present a broad overview of current knowledge on the structure and functions of the NHRs.

Introduction

The family of transcription factors known as nuclear hormone receptors (NHRs) was so named because many are ligand-activated, and thus mediate the biological effects of hormones or hormone-like substances. These include vitamin D_3, thyroxin, retinoids, sex and adrenal steroids, and certain metabolic ligands such as certain fatty acids and oxysterols. Each ligand type binds to and thereby activates one or a closely related set of a few NHR isoforms. In this context, 'activation' refers to ligand-dependent alteration in the ability of the NHR to affect transcription. All members of the NHR family are recognizable due to similarities of certain modular domains, the closest link being provided by the DNA-binding domain (DBD). Including isoforms and depending on species, 100 or more receptor proteins with modular structures that classify them as NHRs may exist, forming the largest family of structurally related metazoan transcription factors, the nuclear receptor superfamily. Many NHRs remain orphan receptors; that is, they have no known ligands as yet. It seems clear now that many NHRs influence transcription in the absence of a ligand. In some cases, that influence is altered when its ligand binds the NHR. Other NHRs may regulate transcription without the requirement for a ligand. Each type of NHR acts to regulate expression of a set of specific genes related to metabolism, development or reproduction. By molecular genetics approaches, it was found that certain regions in the proteins of this group served particular functions, and that these regions could even be transferred among the family members, to make hybrid-function molecules. Thus the concept arose that these receptor proteins comprise semi-autonomous domains or modular structures. As more has been learned, it has become clear that there are limits to this domain concept, since domains may overlap, serve multiple functions and influence one another. Nevertheless, the basic presence and function of the primary domains remains undisputed.

Although the structural organization of NHRs into their three major functional domains, N-terminal transactivation domain (NTD), DBD and C-terminal, ligand-binding domain (LBD) is well characterized (Figure 1A), precisely how transcription is regulated by NHRs is unknown with any degree of rigor. The general outline of how NHRs regulate transcription is known, however. The unliganded receptor may be found in a steady state in both nuclear and cytoplasmic compartments. The proportion in each varies with receptor type, although most NHRs are to be found largely in the nuclear compartment, as the family name implies. For the ligand-activated group, activation frees the receptor to search for and bind with high affinity to specific DNA

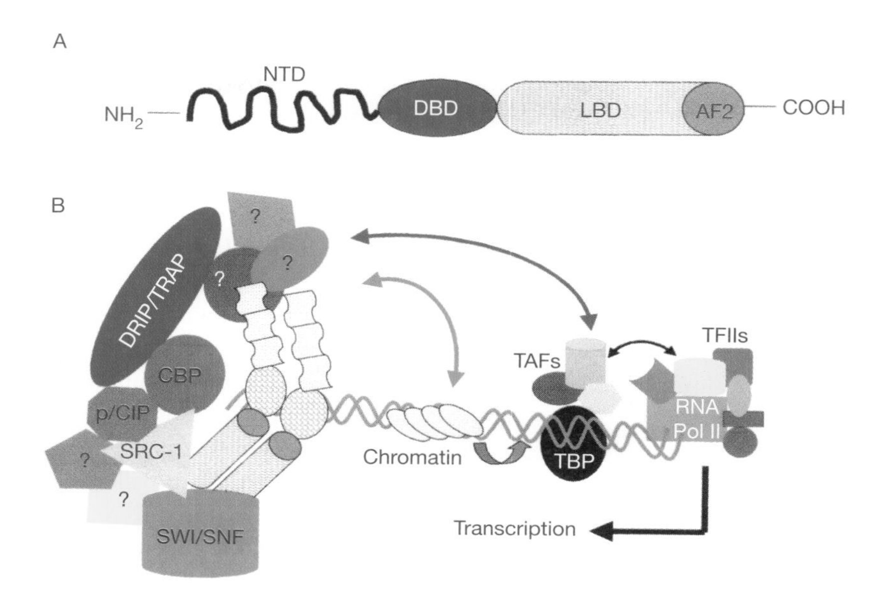

Figure 1. Model for NHR–cofactor complex action on transcription
(**A**) A general topology of the NHRs, showing three principal functional domains. The DBD is the most conserved domain. Its globular structure is responsible for site-specific DNA binding at its response element and some important protein–protein interactions. The second most conserved domain, LBD, is a 12–13-helix globule responsible for ligand binding and aspects of homo- and/or hetero-dimerization. It contains a ligand-dependent transcriptional activation region (AF2). In some cases it may be in the active conformation without a ligand being bound. The diagram shown represents an active conformation. Among members of t he superfamily, the NTD is the least conserved, both in size and primary sequence. When present, the powerful transactivation region AF1 lies within this domain. This region is rich in acidic amino acids, and appears natively to be mostly unstructured. (**B**) A model for the regulation of transcription by NHR–cofactor assemblies. Both AF1 and AF2 regions recruit certain specific cofactors. A bridge is formed between AF1 and AF2 through these and/or other cofactor(s). AF1 and AF2 can also interact directly. The cofactor(s) bound are determined by their levels in particular cell types and by the state of the folded structures of the NHR. The complex alters local chromatin structure (red arrow), e.g. by catalysing histone acetylation or deacetylation, and affects the stabilization of the transcription pre-initiation complex [TATA-box-binding protein (TBP)–TBP-associated factors (TAFs)–RNA polymerase II (RNA Pol II)]. The receptor complex, bound to DNA enhancer sites, thus recruits and regulates polymerase II via accumulations of specific proteins (blue arrow), which make a functional bridge between the receptor and polymerase II. The activity of kinases and phosphatases regulating signalling pathways also contribute to this process by altering the state of phosphorylation of both receptor and cofactors (not shown). The receptor–cofactor assembly may also interact directly with the basal transcription machinery at the TBP/TATA box to regulate transcription. Different colours and shapes show the cofactor proteins. The question marks refer to factors, which may vary depending upon the cell type, ligand and promoter. Precise relationships and site of these cofactor proteins are not accurate and may differ in different NHRs. CBP, cAMP-response-element-binding protein (CREB)-binding protein; TFII, transcription factor II; DRIP/TRAP, vitamin D receptor-interacting protein/ thyroid receptor-associated protein; p/CIP, p300/CBP-interacting protein; SWI/SNF, switch/sucrose non-fermenting; SRC-1, steroid receptor co-activator 1.

sequences (response elements) and protein partners [1]. These presumably bind and release DNA non-specifically until the proper DNA sequence is reached, at which the higher affinity binding somehow controls transcription. At such specific 'response elements', co-operative binding of homo- or hetero-meric NHR dimers localizes the receptor and promotes further protein–protein interactions. These interactions are responsible for changes in transcription of the specific genes specified by the response element. Several mechanisms may be involved, each involving certain protein–protein contacts. The NHR may contact the large, multiprotein primary initiation complex directly, or may influence its functions indirectly, by binding a variety of proteins known as co-activators and co-repressors. These proteins may be able to bind members of the primary transcription complex and/or alter chromatin structure by acetylating or deacetylating histones in the neighbourhood of the cognate response element to which their tethering NHR is bound (Figure 1B). So far as it goes, this model serves well, but it fails to explain explicitly how the relatively small number of different NHRs can achieve the high level of specificity required to regulate the complex pattern of genes affected by NHRs. At present, the general assumption is that the complexes involving various cofactor proteins must do so.

The ability of a small molecule (ligand) to switch each ligand-activatable NHR from its inactive to its active form, thereby causing it to carry signals to regulate specific genes, has facilitated biochemical and biophysical approaches tremendously towards elucidating the mechanism of action of NHRs. Several new discoveries in recent years have added new excitement to the study of NHR function. Among these, the discovery that specific ligands can act as agonists or antagonists to varying degrees in different tissues has completely changed the thinking about how ligands and NHRs function. Structural knowledge of some NHR domains has only begun to explain this phenomenon. Thus, while the general mechanism of how NHRs regulate transcription (discussed above), is useful, it fails to explain several important facts. How do specific ligands within a class act to cause differing NHR functions in various tissues? How do NHRs pass the message from ligand to gene? Why do NHRs possess more than one transactivation region, and what causes these regions often to work in a synergistic manner? What are kinetics of NHR–chromatin interactions? These and other fundamental questions need to be answered in order to understand the full functioning of NHRs as transcription factors. In part, the answers are still missing because little is known about significant aspects of NHR structure, knowledge that is needed to provide the basis for critical specific interactions with other transcription factors. The remainder of this chapter outlines briefly the structural information available for the three major domains found in the NHRs.

NTD

Although detailed structural information is available for several NHR individual domains, expressed as recombinant peptides, no structure of a full-length NHR or of an NTD has been published. In terms of sequence homology, the NTDs are the most poorly conserved regions within the family. NTDs are of widely varying size and sequence. They are often involved in transcriptional regulation within specific cell types. In many NHRs this domain is highly immunogenic, suggesting that it is open to aqueous solvents, and it contains several known phosphorylation sites. One or more powerful transactivation regions are located within the longer NTDs, which, when combined with a DBD in the absence of LBD, can act constitutively (without a ligand) to activate transcription from genes preceded by the proper response element.

In some members of the NHR family, the NTD contains a major transactivation domain (AF1), and some NHRs have two or three AF regions in the NTD. When AF1 domains have been expressed independently as recombinant peptides, they have shown little specific structure, existing as very large collections of conformers that together register as a random coil. Consequently, until recently, experiments attempting to utilize the single-domain expression approach to obtain structural information of the AF1 domain have not been successful. Predictive algorithms for secondary structure suggest that certain AF1s have the capacity to form limited regions of α-helix. At least in the case of glucocorticoid receptor (GR), it has been reported that in the presence of the strong α-helix-promoting agent trifluoroethanol, recombinant AF1 peptide acquires helical segments [2]. Independent experiments showed that helix-breaking double proline substitutions in the potential helical regions interfered with the transactivation function of holo GR containing them. However, due to the high diversity in the length and amino acid sequence of the NTDs, there is as yet no general rule that predicts the structure of all of them.

Other data suggested weak AF1 interactions with certain other proteins relevant to transcription initiation [3]. It is generally presumed that AF1 must bind, directly or via co-activators/co-repressors, to some part(s) of the primary transcription machinery. In the absence of data for a three-dimensional structure of the NTD, several theories have been advanced to explain this. Based on observations that negatively charged areas are important for the transactivation function in the case of some transcription factors, it was proposed that, without having any definite structure, AF1 acts by providing a cloud of negatively charged amino acids. However, mutational analysis of the GR suggests that this is not the case. Perhaps more intellectually satisfying are suggestions that an induced conformation or set of conformations occurs in AF1 in order for it to carry out its transcription function. This kind of induced fit has been reported in otherwise unstructured activation domains of some transcription factors ([4] and references therein). The question is, what causes these conformations? Induced fit could occur when the AF1 domain encounters its proper binding partner(s), or AF1 could exist as an equilibrium between a large proportion of unstructured forms and a small proportion of forms properly structured for functional protein–protein interactions. In the latter case, shifting the equilibrium towards properly structured AF1 conformer(s) would favour binding to the co-regulatory proteins. Alternatively, by the induced-fit model, initial non-specific AF1–co-regulator interactions could result in AF1 becoming properly folded as a heteromeric complex with the protein-binding partner. The alternatives are thermodynamically equivalent [5].

The question for the equilibrium model is what conditions might cause AF1 to acquire the correct conformation? Recent data on the NTD AFs of the GR, androgen receptor (AR), progesterone receptor (PR) and oestrogen receptor (ER) indicate that inter- and intra-molecular forces could be responsible for AF reaching the desired conformation ([5–7] and references therein). Exposing purified recombinant AF1 protein to the osmolyte trimethylamine *N*-oxide (TMAO) results in the domain folding co-operatively into a three-dimensional structure [5,6]. Osmolytes are naturally occurring small compounds that protect proteins in their naturally folded and functional states, and favour such folding from the denatured state due to their solvophobic effects on the peptide backbone ([7] and references therein). The protein backbone comprises the most numerous functional groups in proteins, hence osmolyte-induced conformations are driven by very strong forces, and evidence from several systems studied thus far indicates that native folded functional protein species result ([7] and references therein). Well-known cellular osmolytes include TMAO, glycerol, sorbitol, arabitol, glycine, alanine, proline, betaine and sucrose. They may be found at quite high concentrations in some organisms and organs.

However the physical and functional contacts between AF1 and specific co-regulatory proteins are attained, conditional folding of AF1 appears to be important for these interactions. To give a specific example, the folded AF1 domain of the GR should bind a limited set of specific partners, such as cAMP-response-element-binding protein (CREB)-binding protein (CBP), Ada2 and TATA-box-binding protein (TBP). Weak GR AF1 binding to these was demonstrated in the absence of osmolytes, consistent with the equilibrium. In the presence of the osmolyte TMAO, the interactions are significantly increased, and additionally binding of steroid receptor co-activator 1 (SRC-1) was observed [5]. These findings are consistent with the osmolyte shifting the equilibrium between unstructured and structured AF1 states in favour of the latter, which can then interact efficiently with other binding partners. Similar results have been obtained for the AR [6]. This possibility will become clearer when the three-dimensional structures of some AF1s become available. One interesting point will be to see whether the various AF1s, which differ so considerably in primary sequence, differ or share similarities in three-dimensional structure.

The NTD contains several regions at which post-translational modifications occur. A number of amino acids in the NTD are found to be phosphorylated to varying extents in NHRs isolated from cells or tissues. The effects of site-specific phosphorylations in the NTD on their transcription activating functions vary with particular NHRs [8]. The effects of NTD phosphorylation on AF1 folding have not been investigated in depth. Another important modification of NHR NTDs is sumoylation. At least some NHRs have the SUMO (small ubiquitin-related modifier) peptide added to a lysine residue in their NTDs, at a position between AF1 and the DBD. The effect on structure is unknown, but sumoylation may affect transcriptional transactivation efficiency and possibly nuclear localization [9]. In summary, although as yet no unified structure of an NHR NTD is known, a plausible functional model has been developed. The activation function regions of the NTDs seem to be among the large fraction of protein domains that are natively unfolded ([7] and references therein). The NTD AFs probably undergo conditional folding under the influences of DNA binding by the DBD, AF interactions with co-activators, co-repressors and other proteins or other NHR domains, and possibly the intracellular concentrations of osmolytes.

DBD

Structurally and with respect to primary sequence, the DBD is the most conserved region among the members of NHR superfamily. It contains eight highly conserved cysteine residues, found in two groups of four. Each group is involved in the tetrahedral co-ordination of a single Zn atom. Although these groupings are often referred to casually as zinc fingers, they do not form the three-dimensional structures seen in classic zinc-finger proteins. The three-dimensional structures of the DBD of several NHR members have been solved and discussed extensively ([4] and references therein); therefore only a brief description will be given here. The data from recombinant DBDs indicate that this domain is folded into a globular shape, containing two helices perpendicular to one another, forming the base of a hydrophobic core. One helix is involved in site-specific DNA binding due to its interaction with certain bases in the major DNA groove of its cognate hormone-response element (HRE). Three or four amino acids within this helix are responsible for the site-specific discrimination of binding. The DBD binds to DNA sequences in the HRE as a dimer, with one subunit binding to the HRE specifically and other one non-specifically. The basic building block of these HREs is a receptor-type-specific five- or six-nucleotide sequence comprising a 'half-site'. The classic HRE is composed of a pair of these pentamers/hexamers. The number of non-specific nucleotides providing spacing between them, and their orientation to one another (direct or inverted repeats), makes for a considerable degree of receptor-binding specificity. For some NHRs the DBD–HRE binding is co-operative, with the first specific interaction followed by a sharp increase in affinity as a second DBD monomer is added. Other NHRs, e.g. the ER, appear to dimerize prior to HRE binding. For the steroid NHR subclass, the loop formed between the first two cysteine residues of the second zinc finger provides the dimerization surface. These NHRs dimerize 'head to head' and the base sequences of their HREs are appropriately palindromic. Other classes of NHRs make heterodimers at their HREs. Both the HREs and the protein dimers are head to tail. Prime examples are the thyroid receptor and retinoid receptors, in which the retinoid X receptor heterodimerizes with a thyroid receptor or a retinoid A receptor [10].

The standard model of NHR–DNA binding holds that the effect of the high-affinity binding of receptor to its cognate HRE is to tether the receptor at relevant sites on the genome. Recent data suggest that site-specific DNA binding is a more active event. Site-specific binding of a two-domain recombinant GR to a glucocorticoid response element led to intramolecular signalling in the receptor, causing a conformational change in its AF1 domain [11]. Intramolecular signalling upon site-specific DNA binding also has been shown with the PR. When a two-domain form of the progesterone receptor (NTD+DBD) is expressed, some structure exists in its AF1 domain. DNA binding, however, stabilizes the structure of the region N-terminal to the DBD [12]. Together, these results with two NHRs support the hypothesis that binding to an HRE is not simply a tethering event, as traditional models have held, but a step that initiates intramolecular signals, leading to formation of additional structure in AF1 and other parts of the NTD.

LBD

With respect to amino acid sequence, the LBD is the second most conserved domain of the NHRs, and it is known to participate in several actions of NHRs. The solved three-dimensional structures of several LBDs show great similarity. This domain contains the site for ligand binding, is involved in homo- and/or hetero-dimerization, and transcriptional activation ([4] and references therein). Ligand binding brings conformational changes in the receptor that controls the above-mentioned properties and thus influences gene regulation. Although no extensive evidence is available as to whether these ligand-dependent conformational changes in the LBD cause alterations in other domains, it seems likely that they are affected as well. From crystal structural data on the LBDs of several NHRs, it has been shown that this domain consists of about 12 helices folded into a globular structure. Three sets of helices form the sides and top of the globule, making a central pocket, the ligand-binding site. This overall folding appears to be universally conserved among the NHRs, though as would be expected, the exact dimensions of the pocket vary to suit the ligand(s) specific for each class of NHR ([4] and references therein). Molecular genetics had shown that near the C-terminus of the LBD of several of these hormone-binding receptors lies a short amino acid sequence, known as AF2, important for activating gene transcription ([4] and references therein). Although small, the AF2 transactivation subdomain is very important for the transactivation activity of the NHRs. AF2 binds a number of co-activator and co-repressor proteins via their Leu-Xaa-Xaa-Leu-Leu motifs, and, as mentioned above, these proteins apparently act as molecular bridges between the receptor and the fundamental transcription-initiation complex of proteins. Several co-activators and co-repressors also carry out important enzymic modifications of chromatin [13].

The most C-terminal helix (usually H12) contains the sequence critical to the AF2 transcription-activating function, as it affects the LBD surface upon which much co-activator/co-repressor binding depends. H12 is believed to change position upon ligand binding, flipping from an 'open' position to one closed over the bound ligand. The result is that the LBD now presents a surface favourable for binding the co-activators or co-repressors. At least some ligands that act as antagonists appear to cause altered movement of H12 to a position that creates an unfavourable surface for co-activator binding [14]. However, some steroid hormones, originally identified as antagonists in test systems, are now known to act as agonists in certain tissues. Several possible explanations are offered for the paradox stemming from these indisputable results. In general, it is thought that the large complex of proteins that assembles with the receptor at its AF1 and AF2 sites varies between tissues. Thus, while the fold caused by a given ligand may present binding of a key co-activator in one tissue (antagonist action), in a different tissue, that fold may allow binding of a different but equally effective co-activator (agonist action). When one adds the possibilities of effects due to NHR interactions with other transcription factors at their (non-NHR) binding sites, and the connections with other signal transduction pathways, many hypothetical schemes present themselves. Regardless of mechanism(s), the empirical fact that ligands may be agonists in one tissue and antagonists in another has caused great excitement, as scientists seek to create ligands with desirable tissue-specific therapeutic effects. Undoubtedly, as more becomes known about the details of each individual receptor and its associated protein complexes, features important for unique ligand- and tissue-specific functions will make themselves clear.

The structural dynamics of the LBD and its AF2 also provide a reasonable hypothetical explanation for the fact that some NHRs seem transcriptionally active without need for a ligand. It may be that such NHRs, e.g. NURR1 (NUR-related factor 1) [15], contain an 'LBD' that inherently folds so as to present a favourable surface for co-activator binding. Orphan receptors often have very small NTDs and thus may rely for function largely on the AF activity found in their 'LBDs'. Other known behaviours of NHRs can be understood in terms of the functional structures of their LBDs. The steroid-binding group of NHRs can be activated *in vitro* without ligand by exposure to high salt concentrations and/or mild heating. This may well be due to such conditions causing AF2 to fold in a conformation favourable for binding co-activators. The retinoid/thyroid receptors are thought to be able to bind their response elements and attract co-repressors in the absence of ligands. Specific ligand binding shifts the position of H12, altering the cofactor-binding surface so as to favour co-activator binding. Thus the receptor shifts from a gene-repressing to a gene-activating function.

NHRs and co-regulatory proteins

It is well established that certain specific co-regulatory proteins interact with members of the NHR superfamily. Various co-regulators are either required for or enhance the NHR effect on target gene transcription. These interactions lead to stabilization of NHRs and the basal transcription machinery at the promoter, and to covalent modification of histones, which presumably facilitates or inhibits entry of the transcription complex of proteins to the DNA. Thus these co-regulatory proteins appear to function by either remodelling chromatin structure and/or acting as adapter molecules between NHR and components of the basal transcriptional apparatus. Studies from several laboratories have led to the identification of a series of nuclear-receptor-interacting co-regulatory proteins ([13] and references therein). Most of these are reported to interact with either the AF1 or the AF2 domain, or both. A few other proteins, such as c-Jun, nuclear factor κB, GT198 and some signal tranducers and activators of transcription (STATs) interact with the DBD. Other proteins are recruited to the co-regulators that bind the NHR directly, so that a large NHR–co-regulator complex is assembled. Two classes of co-activator complex are recruited either simultaneously or combinatorially. One set promotes the nucleosomal remodelling required for transcriptional activation or repression, and another one forms a direct bridge to the transcriptional machinery apparatus. The structural nature of these large molecular complexes remains a challenge for the future.

Although in most experimental systems either AF1 or AF2 alone is capable of regulating transcription to some extent, full transcriptional activation by NHRs usually requires functional synergy between the two transactivation function regions ([7] and references therein). Recent data show that for the GR at least, the nature of the DNA template is critical to the effect seen. For induction from a promoter in a natural chromatin background, AF1 was quantitatively more important than AF2 [16]. In a natural holoreceptor it is likely that a physical intramolecular association occurs between the NTD and the LBD. This has been supported by studies with the PR-B isoform, in which a ligand-dependent, direct interaction between the N-terminal and C-terminal domains has been shown [17]. Such an interaction takes place only when an agonist is bound, and antagonist binding prevents it, suggesting a conformational change in the agonist-bound LBD that suits it for this interaction. Furthermore, ubiquitous co-activators such as SRC-1 and p300/CBP are not required for this physical association; thus it seems that the LBD–NTD interaction may be a direct one. AR NTD and LBD interactions are also well documented [18].

Bridging proteins between the two NHR domains may come into play in some circumstances. It has been reported that through different regions, SRC-1 is able to interact with both the AF1 and AF2 regions of several steroid receptors. Co-operativity observed between AF1 and AF2 regions appears to be assisted as well by other cofactor proteins, such as TIF2 [19]. Another co-activator complex, DRIP/TRAP (vitamin D receptor-interacting protein/thyroid receptor/associated protein), also has been reported to interact with both AF1 and AF2 regions in several NHRs [20]. Observations of synergism between AF1 and AF2 are also observed involving p300/CBP and the ER [21]. One report indicates that SRC-1 interaction with the AF1 region of the ER requires the putative α-helical core of the AF1 region [22], consistent with results from the GR AF1, where it has been shown that conditional folding of the AF1 domain is a prerequisite for its interaction with SRC-1 [5].

Conclusions

In the certainty that protein function depends on structure, studies of NHR structure have been pursued vigorously in recent years. Up to the present, these studies have utilized recombinant proteins representing portions of the holoreceptors. Recombinant proteins encompassing the DBDs and/or LBDs of several NHRs have been shown to fold into globular forms, allowing their structures to be determined. In conjunction with a vast amount of other experimentation in many laboratories, these structures have greatly aided our understanding of how the NHRs interact with their specific DNA-binding sites, dimerize with one another, and respond to the binding of specific ligands. The NTDs of the NHRs have been more difficult to study structurally, because when they are exposed as recombinant proteins they show little structure. Recently, however, methods have been developed that cause NTDs to assume three-dimensional structures, suggesting that these will become more accessible for study.

Analyses short of full structural determinations on recombinant proteins containing two NHR domains have begun to appear. Early data from these suggest that site-specific DNA binding results in acquisition of significant structure in the attached NTD. This is consistent with hints from molecular biological experiments suggesting that particular sequences within DNA-binding sites for NHRs may affect NHR functions. NHRs also affect transcription through protein–protein interactions at non-cognate DNA sites. The structural basis for this type of control over transcription is obscure.

Structures of holo NHRs have yet to be achieved, and for full understanding of their actions, these structures will be necessary. Indeed, well beyond that, we will need to discover the structures of the large, multiprotein complexes that are obviously critical to NHR function.

Summary

- *NHRs are multi-domain proteins that function as transcription factors.*
- *NHR family members share identity owing to the structural similarity of their DBDs.*

- *Classic NHRs contain three major domains: the NTD, the DBD and the LBD. (i) NTDs contain one or more regions that are important in activating transcription, e.g. AF1. (ii) LBDs also contain a transcription-activating region, AF2. (iii) DBDs bind to specific DNA sequences, or response elements. (iv) Upon binding with their specific ligand, such as a hormone or similar signalling molecule, classic NHRs become active as transcription factors.*
- *Non-classic NHRs (orphan receptors) are those for which no ligand has been identified.*
- *NHRs affect transcription through interactions with a variety of other proteins — co-activators, co-repressors, parts of the primary transcription complex etc,.*
- *Three-dimensional structures of several NHR DBDs and LBDs, independently expressed as recombinant proteins, have been obtained. (i) The DBD structures are all similar and show how the NHR DBDs interact as dimers with their specific response elements. (ii) The LBD structures reveal a set of conserved helices that provide NHR-unique ligand-binding pockets. Various ligands alter the position of the helix that is critical for AF2 function, and thus alter the surface for interaction with other critical proteins. (iii) No three-dimensional structure is available for an AF1 from an NTD*
- *As recombinant proteins, AF1s are not fully structured. (i) A greater degree of structure in AF1s can be brought about by the use of osmolytes (small molecules that enhance protein folding), by binding an attached DBD to its cognate response element, and/or by binding between the AF1 and one of its protein-binding partners. (ii) It is believed that such mechanisms confer transcription-regulating function on AF1.*

Supported by grants from the National Institutes of Health (NIDDK 1RO1-DK-58829; to R.K. and E.B.T.) and the Muscular Dystrophy Association (to R.K.).

References

1. Beato, M., Herrlich, P. & Schutz, G. (1995) Steroid hormone receptors: many actors in search of a plot. *Cell* **83**, 851–857
2. Dahlman-Wright, K., Baumann, H., McEwan, I.J., Almlof, T., Wright, A.P.H., Gustafsson, J.-Å. & Hard, T. (1995) Structural characterization of a minimal functional transactivation domain from the human glucocorticoid receptor. *Proc. Natl. Acad. Sci. U.S.A.* **92**, 1699–1703
3. Almlof, T., Wallberg, A.E., Gustafsson, J.-Å. & Wright, A.P.H. (1998) Role of important hydrophobic amino acids in the interaction between the glucocorticoid receptor tau1-core activation domain and target factors. *Biochemistry* **37**, 9586–9594
4. Kumar, R. & Thompson, E.B. (1999) The structure of the nuclear hormone receptors. *Steroids* **64**, 310–319

5. Kumar, R., Lee, J.C., Bolen, D.W. & Thompson, E.B. (2001) The conformation of the glucocorticoid receptor AF1/tau1 domain induced by osmolyte binds co-regulatory proteins. *J. Biol. Chem.* **276**, 18146–18152
6. Reid, J., Kelly, S.M., Watt, K., Price, N.C. & McEwan, I.J. (2002) Conformational analysis of the androgen receptor amino-terminal domain involved in transactivation. *J. Biol Chem.* **277**, 20079–20086
7. Kumar, R. & Thompson, E.B. (2003) Transactivation functions of the N-terminal domains of nuclear hormone receptors: protein folding and coactivator interactions. *Mol. Endocrinol.* **17**,1–10
8. Orti, E., Bodwell, J.E. & Munck, A. (1992) Phosphorylation of steroid hormone receptors. *Endocr. Rev.* **13**, 105–128
9. Le Drean, Y., Mincheneau, N., Le Goff, P. & Michel, D. (2002) Protection of glucocorticoid receptor transcriptional activity by sumoylation. *Endocrinology* **143**, 3482–3489
10. Perlmann, T., Rangarajan, P.N., Umesono, K. & Evans, R.M. (1993) Determinants for selective RAR and TR recognition of direct repeat HREs. *Genes Dev.* **7**, 1411–1422
11. Kumar, R., Baskakov, I.V., Srinivasan, G., Bolen, D.W., Lee, J.C. & Thompson, E.B. (1999) Interdomain signaling in a two-domain fragment of the human glucocorticoid receptor. *J. Biol. Chem.* **274**, 24737–24741
12. Bain, D.L., Franden, M.A., McManaman, J.L., Takimoto, G.S. & Horwitz, K.B. (2000) The N-terminal domain of the progesterone A-receptor structural analysis and the influence of the DNA binding domain. *J. Biol. Chem.* **275**, 7313–7320
13. McKenna, N.J., Lanz, R.B. & O'Malley, B.W. (1999) Nuclear receptor coregulators: cellular and molecular biology. *Endocr. Rev.* **20**, 321–344
14. Brzozowski, A.M., Piki, A.C.W., Dauter, Z., Hubbard, R.E., Bonn, T., Engstrom, O., Ohman, L., Greene, G.L., Gustafsson, J.-Å. & Carlquist, M. (1997) Molecular basis of agonism and antagonism in the oestrogen receptor. *Nature (London)* **389**, 53–58
15. Wang, Z., Benoit, G., Liu, J., Prasad, S., Aarnisalo, P., Liu, X., Xu, H., Walker, N.P.C. & Perlmann, T. (2003) Structure and function of Nurr1 identifies a class of ligand-independent nuclear receptors. *Nature (London)* **423**, 555–561
16. Keeton, E.K., Fletcher, T.M., Baumann, C.T., Hager, G.L. & Smith, C.L. (2002) Glucocorticoid receptor domain requirements for chromatin remodeling and transcriptional activation of the mouse mammary tumor virus promoter in different nucleoprotein contexts. *J. Biol. Chem.* **277**, 28247–28255
17. Tetel, M.J., Giangrande, P.H., Leonhardt, S.A., McDonnell, D.P. & Edwards, D.P. (1999) Hormone-dependent interaction between the amino- and carboxyl-terminal domains of progesterone receptor *in vitro* and *in vivo*. *Mol. Endocrinol.* **13**, 910–924
18. He, B., Kemppainen, J.A., Voegel, J.J., Gronemeyer, H. & Wilson, E.M. (2002) Activation function-2 in the human androgen receptor ligand binding domain mediates interdomain communication with NH_2-terminal domain. *J. Biol. Chem.* **274**, 37219–37225
19. Bommer, M., Benecke, A., Gronemeyer, H. & Rochette-Egly, C. (2002) TIF2 mediates the synergy between RARa1 activation functions AF-1 and AF-2. *J. Biol. Chem.* **277**, 37961–37966
20. Hittelman, A.B., Burakov, D., Iniguez-Lluhi, J.A., Freedman, L.P. & Garabedian, M.J. (1999) Differential regulation of glucocorticoid receptor transcriptional activation via AF-1-associated proteins. *EMBO J.* **18**, 5380–5388
21. Kobayashi, Y., Kitamotom T., Masuhiro, Y., Watanabe, M., Kase, T., Metzger, D., Yanagisawa, J. & Kato, S. (2000) p300 mediates functional synergism between AF-1 and AF-2 of estrogen receptor α and β by interacting directly with the N-terminal A/B domains. *J. Biol. Chem.* **275**, 15645–15651
22. Metivier, R., Penot, G., Flouriot, G. & Pakdel, F. (2001) Synergism between ERa transactivation function 1 (AF-1) and AF-2 mediated by steroid receptor coactivator protein-1: requirement for the AF-1 α-helical core and for a direct interaction between the N- and C-terminal domains. *Mol. Endocrinol.* **15**, 1953–1970

4

Role of molecular chaperones in steroid receptor action

William B. Pratt[1], Mario D. Galigniana, Yoshihiro Morishima and Patrick J. M. Murphy

Department of Pharmacology, The University of Michigan Medical School, Ann Arbor, MI 48109, U.S.A

Abstract

Unliganded steroid receptors are assembled into heterocomplexes with heat-shock protein (hsp) 90 by a multiprotein chaperone machinery. In addition to binding the receptors at the chaperone site, hsp90 binds cofactors at other sites that are part of the assembly machinery, as well as immunophilins that connect the assembled receptor–hsp90 heterocomplexes to a protein trafficking pathway. The hsp90-/hsp70-based chaperone machinery interacts with the unliganded glucocorticoid receptor to open the steroid-binding cleft to access by a steroid, and the machinery interacts in very dynamic fashion with the liganded, transformed receptor to facilitate its translocation along microtubular highways to the nucleus. In the nucleus, the chaperone machinery interacts with the receptor in transcriptional regulatory complexes after hormone dissociation to release the receptor and terminate transcriptional activation. By forming heterocomplexes with hsp90, the chaperone machinery stabilizes the receptor to degradation by the ubiquitin–proteasome pathway of proteolysis.

Introduction

Heat-shock protein (hsp) 90 is an abundant (1–2% of cytosolic protein), conserved and essential protein of eukaryotic cells. Although it is an hsp, its

[1]*To whom correspondence should be addressed (e-mail annat@umich.edu).*

manufacture being increased in stressed cells, it performs a variety of housekeeping functions in unstressed cells. For the steroid receptors, hsp90 functions as a cradle-to-grave chaperone, in that it regulates proper receptor folding, trafficking, transcriptional activation and turnover [1]. Some 20 years ago, hsp90 was shown to be associated with the oncogenic protein tyrosine kinase v-Src, and it is now known to bind to and regulate the function of approx. 100 transcription factors and protein kinases involved in signal transduction, a process of identification that was greatly aided by the hsp90 inhibitor geldanamycin [2]. Of these many 'client' proteins for hsp90, the assembly of steroid receptor–hsp90 complexes is the best defined, and most studies of hsp90's effects on receptor function *in vivo* have focused on the glucocorticoid receptor (GR). In this chapter, we will use the GR to illustrate the effects of hsp90 and other molecular chaperones (e.g. hsp70) on steroid receptors. The same mechanism that we describe here for assembly of GR–hsp90 heterocomplexes applies to the assembly of hsp90 complexes with other transcription factors and with protein kinases involved in a variety of signalling pathways [2].

Hsp90 acts on the ligand-binding domain (LBD) of steroid receptors

Although hsp90 and hsp70 may function individually as chaperones that bind to exposed hydrophobic amino acids in denatured or partially denatured proteins and through rounds of binding and release facilitate protein renaturation, this is not the way that they function on steroid receptors. The binding of hsp90 to the receptors is the result of the co-ordinated action of several proteins that function together as an hsp90-/hsp70-based chaperone machinery for the ATP-dependent assembly of receptor–hsp90 heterocomplexes. These complexes are formed specifically with the LBD in the C-terminal one-third of the receptor (Figure 1). For the GR and some of the other steroid receptors, hsp90 must be bound to the LBD for it to have high-affinity steroid-binding activity. Removal of hsp90 from the GR (with salt, for example) is accompanied by immediate loss of steroid-binding activity, and the LBD can be returned to the steroid-binding state by the chaperone machinery. There is no evidence that the hsp90-free GR is in any way denatured; rather, the hydrophobic steroid-binding cleft appears to be collapsed such that the LBD is in a native, minimal-energy conformation. Energy in the form of ATP is then required for the hsp90-/hsp70-based chaperone machinery to open up the cleft during GR–hsp90 heterocomplex assembly, regenerating steroid-binding activity. Thus when hsp90 and hsp70 operate together as part of the chaperone machinery, they open a hydrophobic cleft in a properly folded state of the GR, rather than promoting the refolding of a partially denatured GR, as occurs classically in chaperone-mediated events. Once the cleft is open, it remains open as long as the LBD is maintained in a metastable complex with hsp90.

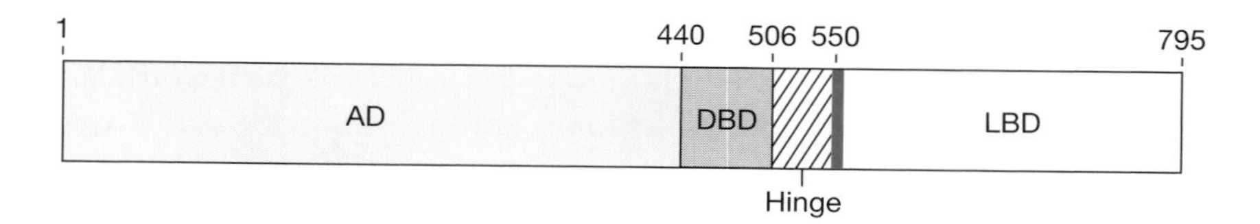

Figure 1. Domain structure of the GR
The borders of four domains of the rat GR are defined: the activation domain (AD), involved in co-factor interaction; the DNA-binding domain (DBD); the hinge domain, which contains a nuclear localization signal and a binding site for FKBP52; and the LBD. The short segment in blue at the extreme N-terminus of the LBD lies at the lip of the steroid-binding cleft and is required for hsp90 binding to the GR.

Subsequent binding of steroid deep within the cleft facilitates a temperature-dependent collapse, or closing, of the cleft with loss of the GR's ability to form 'persistent' complexes with hsp90. This liganded form of the GR is said to be 'transformed' or 'activated' because it can now move to the nucleus and be transcriptionally active [1]. Although the transformed, liganded GR can interact in very dynamic fashion with the chaperone machinery, the persistent heterocomplexes with hsp90 that are typical of client protein–hsp90 complexes in general can only be formed again after the steroid has been released from its hydrophobic nest in the interior of the LBD. This ability of the steroid to regulate the interaction of the receptor with the chaperone machinery is the key initial event in signal transduction by steroids.

Hsp90 acts at the opening of the steroid-binding cleft

The study of the interaction of the chaperone machinery with the GR may have given us some insight as to how hsp90 can form complexes with so many signalling proteins regardless of their amino acid sequence and without a requirement that the protein be in a non-native state. There is a very focal site of attack for the chaperone machinery on the surface of the GR that lies at the N-terminus of the LBD. Hsp90 binding requires the presence of a short segment of the LBD that has no defined amino acid composition (Figure 1), and this segment lies at the rim of the steroid-binding cleft [3]. Virtually all soluble proteins in native conformation have regions where the hydrophobic surfaces of the protein interior merge with their hydrophilic exterior. Such hydrophobic cleft openings may be the general topological feature recognized by hsp70, which is the first component of the chaperone machinery to interact with the receptor [4]. We will return to this cleft-opening model below when we consider the choice between receptor stabilization by hsp90 versus receptor degradation via the ubiquitin–proteasome pathway.

The importance of the short segment at the rim of the ligand-binding cleft for both hsp90 binding and regulation of receptor function *in vivo* is highlighted by observations with a GR–retinoic acid receptor (RAR) chimaera [5]. Although the RAR requires the chaperone machinery to achieve its original steroid-binding state, it does not form persistent complexes with hsp90, and it does not remain in the cytoplasm of hormone-free cells, as does the GR. When a fragment of the GR from the N-terminus through the dark blue segment shown in Figure 1 was fused to the complementary RAR LBD fragment extending to its C-terminus, the chimaera was complexed with hsp90 and underwent cytoplasmic–nuclear translocation in response to retinoic acid.

The LBD is a transferable regulatory unit

It has been known for many years that fusion of a steroid receptor LBD to other transcription factors, such as the viral E1A protein, may yield a chimaera in which transcriptional activity is regulated by steroid binding [6]. This ability of the LBD to function as a moveable 'regulatory cassette' is now known to be due to the transfer of steroid-regulated hsp90 binding on to an otherwise constitutively acting protein. The cleft hypothesis gives us a general basis for understanding how such a regulatory system may have evolved. It is likely that the interaction of the chaperone machinery with the regions where clefts merge with the surfaces of most native proteins is a highly dynamic process. We will see later how such dynamic interactions are useful for attaching the liganded receptor to a system for rapid protein movement. The evolution of more persistent hsp90 attachment to the now partially opened cleft allowed the steroid to enter the hydrophobic cleft and regulate the hsp90 interaction with the receptor. In other cases, hsp90-dependent opening of hydrophobic clefts may facilitate the entry of hydrophobic cofactors, such as haem or flavin, into the interior of an enzyme. This is the case with the more dynamic interaction of hsp90 with neuronal nitric oxide synthase, where the chaperone machinery acts on the apo-enzyme to promote haem entry, which in turn permits homodimerization and activation of catalytic activity [7]. In many signal transduction systems, a persistent interaction of hsp90 with a protein, or proteins, in the signalling pathway stabilizes the protein to rapid degradation by the ubiquitin–proteasome pathway. Thus the chaperone machinery serves to maintain the abundance of a key protein, usually a protein kinase, such that normal signal transduction occurs. In this case, cleft opening may not be critical for permitting entry of a ligand, but we will discuss below how interaction of the machinery with the cleft opening is likely to be critical for determining whether an hsp90-regulated protein will be stabilized or degraded.

Assembly of receptor–hsp90 heterocomplexes

Although receptor–hsp90 heterocomplexes could readily be isolated from cell cytosols by immunoprecipitation of unliganded receptors, initial attempts to form them under cell-free conditions failed. Purified hsp90 does not bind to purified receptor, and it was not until receptors were translated in reticulocyte lysate that cell-free assembly was achieved. A system was then set up in which an unliganded receptor was immunoadsorbed, and the receptor was stripped of its bound hsp90 by washing the immune pellets with a salt solution. In the case of the GR, we now have an aporeceptor without steroid-binding activity. When this immunopurified aporeceptor is incubated with rabbit reticulocyte lysate, the GR is assembled into a heterocomplex with hsp90, with simultaneous regeneration of steroid-binding activity. Once this cell-free assembly system was available, the major components of the assembly system were identified. Receptor–hsp90 assembly is now carried out with a combination of five purified proteins – hsp90, hsp70, Hop (hsp organizer protein), hsp40, p23 – that comprises a minimal system for efficient assembly of signalling protein–hsp90 heterocomplexes in general [2].

The master chaperones: hsp90 and hsp70

Hsp90 and hsp70 are both chaperones possessing nucleotide-binding sites that regulate their conformations. In each case, the ATP-bound conformation has a low affinity for binding hydrophobic peptide, and ATP hydrolysis resulting from the intrinsic ATPase activity of the chaperone is accompanied by a conformational change to a state with high affinity for binding hydrophobic peptide [8,9]. The ATPase activity of both proteins is essential for their biological function and is regulated by co-chaperones. Hsp70 is essential for assembly of receptor–hsp90 heterocomplexes. Although the two proteins alone do not assemble GR–hsp90 heterocomplexes that are persistent, they are sufficient for transient opening of the steroid-binding cleft [10]. The hydrolysis of ATP by hsp70 is potassium-dependent, and GR–hsp90 heterocomplex assembly is K^+-dependent, as well as MgATP-dependent. Hsp90 contains a dimerization domain at its C-terminus, it functions as a dimer, and there is one molecule of receptor bound to two molecules of hsp90 in the final heterocomplex. Hsp90 is a member of a small family of proteins, the GHKL (gyrase, hsp90, histidine kinase, MutL) family, which possess a unique binding pocket for ATP. The hsp90 inhibitors geldanamycin and radicicol bind to this N-terminal nucleotide-binding site and prevent hsp90 from assuming its ATP-dependent conformation, thus blocking receptor–hsp90 heterocomplex assembly and opening of the steroid-binding cleft.

The hsp90-/hsp70-based chaperone machinery

When hsp70 and hsp90 act on the GR in reticulocyte lysate they act in concert with several other proteins as a machinery in which Hop brings the two master chaperones together, as shown in Figure 2. Hop contains two tetratricopeptide repeat (TPR) domains that bind independently to hsp90 and hsp70. TPR domains are 34-amino-acid repeats that function as protein-binding regions, and hsp90 has a TPR acceptor site at its C-terminus. Although Hop is not essential for GR–hsp90 heterocomplex assembly, when Hop is present to bring hsp90 and hsp70 together, forming a machinery, the rate of assembly is much faster [10]. Hop does not act solely as a tie rod to bring the master

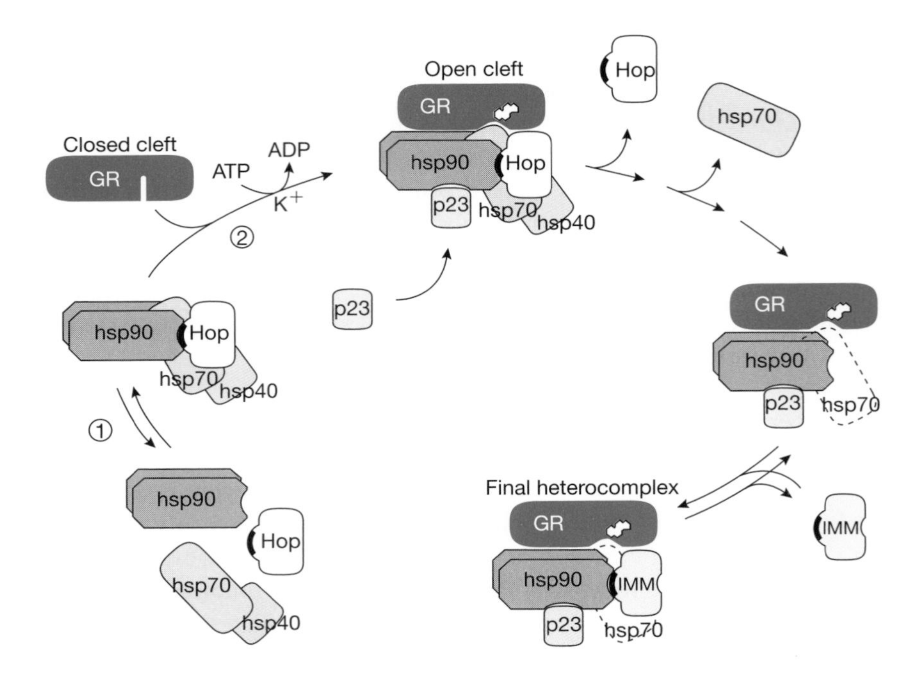

Figure 2. Model of GR–hsp90 heterocomplex assembly
The hsp90-/hsp70-based chaperone machinery converts the GR LBD from a folded conformation in which the steroid-binding cleft is closed and not accessible to hormone into an open-cleft conformation that can be accessed by steroid. Hop binds via independent TPR (tetratricopeptide repeat) domains to hsp90 and hsp70 to form the chaperone machinery. The hsp90–Hop–hsp70–hsp40 machinery contains all the components necessary to carry out the MgATP- and K^{+}-dependent opening of the steroid-binding cleft. When the cleft is open, hsp90 is in its ATP-dependent conformation, which is dynamically stabilized by p23, and Hop is released from hsp90. Thus the complex shown at the end of step 2 represents a composite in which several changes are occurring in dynamic fashion. After the exit of Hop, an immunophilin (IMM) can bind to the single TPR acceptor site on the receptor-bound hsp90 dimer. The hsp90-binding TPR domains of Hop and immunophilins are indicated by the solid black crescents. Much of the hsp70 leaves the intermediate complex during assembly, and the broken line for hsp70 in the final heterocomplex indicates that it is present at substoichiometric levels with respect to the receptor.

chaperones together; it also influences the conformational state and function of each protein [11]. As indicated in Figure 2, the machinery also contains hsp40, a co-chaperone that binds to hsp70 and promotes its ATPase activity. The machinery forms spontaneously on mixing of the purified proteins, and the immunoadsorbed hsp90–Hop–hsp70–hsp40 complex has everything necessary to convert the LBD into the steroid-binding state when it is incubated with the immunoadsorbed GR.

Although all of the Hop and about one-third of the hsp90 in reticulocyte lysate are present in the hsp90–Hop–Hsp70–hsp40 complex [12], it is not known if the preformed machinery interacts with the closed cleft form of the GR, as shown in the simplified assembly model in Figure 2. There may be a requirement that the machinery be assembled on the receptor as the individual components bind in a specific sequence to open the steroid-binding cleft during receptor–hsp90 heterocomplex assembly [13,14].

Assembly is a dynamic process

To achieve an open steroid-binding cleft, the receptor-bound hsp90 must assume a conformation induced by ATP. This is the conformation of hsp90 that recognizes p23, which now binds the receptor-bound hsp90 and stabilizes the GR–hsp90 heterocomplex [15]. p23 is the only member of the five-protein assembly system that does not exist in the machinery (i.e. the hsp90–Hop–hsp70–hsp40 complex), but it is shown in Figure 2 to bind to the GR-bound hsp90 when the other members are present as part of a receptor-bound machinery. This is a diagrammatic oversimplification of what is really a dynamic series of events. As the steroid-binding cleft is being opened, hsp90 achieves its ATP-bound state, Hop dissociates from the receptor-bound hsp90, and p23 binds. Also, a considerable amount of hsp70 dissociates from the GR during the cleft-opening process, leaving a steroid-binding state of the GR that is bound to hsp90, p23 and a variable, but substoichiometric, amount of hsp70 (indicated by the shape with the dashed outline in Figure 2). Although a general sequence of events can be described, the factors that control the exit of Hop and hsp70 are not known.

The product of assembly with the purified five-protein system is a GR–hsp90–p23 complex in which the TPR acceptor site on hsp90 is empty. In reticulocyte lysate or in intact cells where a variety of TPR domain immunophilins are available, the immunophilins compete with each other to occupy the TPR acceptor site. Thus as we will describe, a variety of final heterocomplexes may exist, depending upon the immunophilin that is occupying the TPR acceptor site on hsp90 at any time.

Mechanism of cleft opening

Although the cleft-opening process is indicated as a single step (step 2) in Figure 2, stepwise assembly experiments have begun to separate the process into an ordered series of events shown in Figure 3. Two independent, ATP-requiring steps have been resolved [13].

Priming of the receptor by hsp70

The first step in assembly is the 'priming' of the GR to form a GR–hsp70 complex that then binds Hop and hsp90. In this first step, the immunoadsorbed GR is incubated with hsp70 and hsp40 in the presence of MgATP and K^+. This produces a 'primed' GR–hsp70–hsp40 complex that can be washed free of unbound hsp70 and hsp40, and then incubated with purified Hop, hsp90 and p23 to produce a GR–hsp90 complex with an open steroid-binding cleft that can be accessed by hormone. Hsp40 is not essential for GR

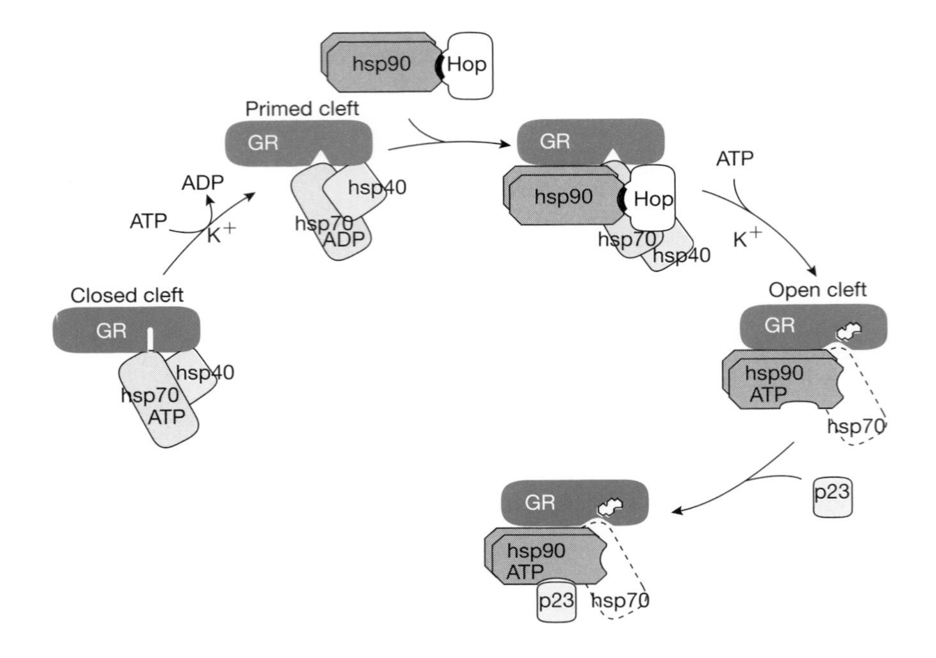

Figure 3. Mechanism of cleft opening by hsp70 and hsp90
The ATP-dependent conformation of hsp70 binds initially to the naked GR, and in an ATP-, K^+- and hsp40-dependent step, a 'primed' GR–hsp70 complex is formed that can bind hsp90. After hsp90 binding, there is a second ATP- and K^+-dependent step (or steps) that is rate-limiting and leads to opening of the steroid-binding cleft to access by a steroid. The GR-bound hsp90 is now in its ATP-dependent conformation and can be bound by p23, which stabilizes the chaperone in that conformation, preventing disassembly of the GR–hsp90 heterocomplex. The hsp40 and Hop components of the five-protein assembly system have been omitted from later steps for simplicity.

priming, but its presence doubles or triples the amount of primed GR–hsp70 complex that is produced. Thus both ATP binding and ATP hydrolysis are required for priming.

The attack on the GR is not only focal at the cleft opening but also limited in the sense that only one or at most two molecules of hsp70 are bound to the GR in the primed complex. The receptor in the primed GR–hsp70 complex cannot bind steroid, but something has happened to the steroid-binding cleft, probably a partial opening of the cleft, such that it can accept hsp90. Hsp70 in its ATP-dependent conformation binds to the naked GR, and if hsp70 is converted first into its ADP-dependent conformation, there is no binding to the GR [4]. After the priming step, the GR-bound hsp70 is in both ADP-bound and ATP-bound states, and there is evidence that once hsp70 has bound to the GR, the chaperone may oscillate back and forth between ATP-bound and ADP-bound configurations during the priming step [16].

Cleft opening by hsp90

After priming, the washed GR–hsp70 complex can bind hsp90 and Hop to form a GR–hsp70–Hop–hsp90 complex. At 30°C, hsp90 binding is rapid and nucleotide-independent. This complex of the GR with the attached chaperone machinery does not yet have steroid-binding activity, but it can be washed free of unbound hsp90 and Hop and then incubated with ATP to yield an open-cleft form of the receptor with steroid-binding activity. Most of the GR-bound hsp70 dissociates as cleft opening occurs, and when the cleft is open, the hsp90 is in the ATP-induced conformation that is recognized by p23. In this series of reactions, both the priming step (approx. 1 min) and the hsp90-binding step (approx. 30 s) are rapid at 30°C, and it is the ATP-dependent opening of the binding cleft (10–15 min) that is rate-limiting for the overall process.

Other co-chaperones

The five-protein system is a minimal system for efficient receptor–hsp90 heterocomplex assembly, and there are additional co-chaperones in this process in cells. For example, a family of hsp90 co-chaperones called Aha (activator of hsp90 ATPase) has recently been identified [17]. The Aha proteins are themselves stress proteins, and they bind to hsp90 and stimulate its inherent ATPase activity, much like DnaJ proteins, such as hsp40, stimulate the ATPase activity of hsp70. Aha1 has been shown to be necessary for hsp90-dependent activation of the protein tyrosine kinase v-Src *in vivo*, and it seems that it must play some role in steroid-receptor regulation by hsp90 *in vivo*, although it may do so only under stress conditions.

BAG-1 (Bcl-2-associated gene product-1), also called RAP46 (46 kDa receptor-associated protein), is a co-chaperone that binds to the ATPase domain of hsp70 and accelerates the release of ADP. BAG-1 is recovered in steroid receptor heterocomplexes, and overexpression of BAG-1 inhibits GR binding to DNA and GR-dependent transactivation. Inhibition of GR-dependent transcription by BAG-1 requires both its binding to hsp70, and BAG-1's ability to bind directly in a non-sequence-specific manner to DNA [18,19]. The hsp90-/hsp70-based chaperone machinery is active in the nucleus as well as the cytoplasm, and BAG-1 is likely to play an important role in terminating transcriptional activation as GR–hsp90 heterocomplexes are formed when hormone levels decline. Another hsp70 co-chaperone found in steroid receptor heterocomplexes is Hip (hsp70-interacting protein). Hip also binds to the ATPase domain of hsp70, where it competes for the binding of BAG-1. It does not affect GR–hsp90 assembly by the five-protein machinery, but Hip and BAG-1 may play opposing regulatory roles on dynamic receptor–hsp90 assembly/disassembly in cells.

Hsp90 and GR function *in vivo*

Of the several effects of the chaperone machinery on GR function *in vivo*, the requirement of hsp90 for steroid-binding activity is the only one that has been studied in the purified system. Other effects of the machinery on receptor trafficking, nuclear receptor cycling and receptor turnover have been studied indirectly by using hsp90 inhibitors, such as geldanamycin and radicicol, or by overexpressing proteins or protein fragments that affect hsp90 function.

Receptor trafficking

The steroid receptors constantly pass into and out of the nuclei of hormone-free cells, and depending upon the receptor and the type of cell, the unliganded receptor at steady state may be predominantly cytoplasmic or nuclear [20]. In most cells, the unliganded GR is in the cytoplasm, and when it becomes bound by steroid it shifts to the nucleus. Normally, this steroid-dependent translocation is rapid ($t_{1/2}$ approx. 4.5 min), but treatment of cells with the hsp90 inhibitor geldanamycin lowers the rate of translocation by an order of magnitude. The rapid, hsp90-dependent movement occurs along cytoskeletal tracts, and the slow, hsp90-independent movement probably reflects diffusion [21]. In neurites, where random diffusion alone does not permit delivery of proteins over long distances, treatment with geldanamycin blocks GR movement. Thus a facilitory action of hsp90 in retrograde protein trafficking in the cell body becomes essential for trafficking in the specialized extensions of the cytoplasm that comprise the axons and dendrites. The binding of steroid deep within the binding pocket triggers the transformation of the receptor LBD to a state where it interacts very dynamically with the chaperone machinery, bringing the receptor into continuous association with and dissociation from hsp90 and the immunophilins, which, in turn, link the receptor complex to the motor protein responsible for retrograde movement to the nucleus.

Cytoplasmic dynein is known to be the motor protein responsible for retrograde movement of vesicles and organelles along microtubular tracks toward the nucleus [22], and the connection between the GR and the motor via the hsp90-bound immunophilins is illustrated in Figure 4. As we have shown in Figure 2, the final GR–hsp90 heterocomplex is formed when a TPR domain immunophilin binds to the TPR acceptor site in the C-terminus of hsp90. The immunophilins are ubiquitous and conserved proteins that bind immunosuppressant drugs, such as FK506, rapamycin and cyclosporin A. All members of the immunophilin family have peptidylprolyl isomerase (PPIase) activity, and they may be divided into two classes: the FK506-binding proteins (FKBPs) are binding proteins for immunosuppressant compounds such as FK506 and rapamycin, and the cyclophilins (CyPs) bind cyclosporin A. The immunosuppressant drugs occupy the PPIase sites on the immunophilins and inhibit *cis–trans* isomerization of petidylprolyl bonds *in vitro*. For the steroid

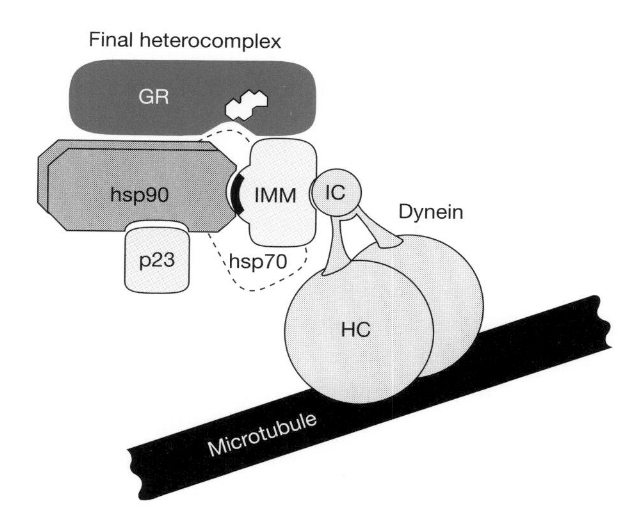

Figure 4. The GR–hsp90–immunophilin heterocomplex attaches to cytoplasmic dynein for retrograde movement along microtubule highways
Cytoplasmic dynein is the motor protein that processes along microtubules in a retrograde movement toward the nucleus. It is a very large multisubunit complex (approx. 1.2 MDa) comprising two heavy chains (HC) that have the processive motor activity, three intermediate chains (IC) that link to cargo, and some light chains that are not shown. The immunophilin (IMM) component of the GR–hsp90 heterocomplex links via its PPIase domain to the intermediate chain of dynein.

receptors, however, PPIase activity does not appear to be important; rather, the PPIase domains act as protein–protein interaction domains to link the receptors to cytoplasmic dynein [23].

Four high-molecular-mass immunophilins with TPR domains have been discovered as components of steroid receptor–hsp90 complexes: FKBP52, FKBP51, CyP40 and PP5, the latter of which contains a protein phosphatase domain in addition to TPR and PPIase homology domains. When GR–hsp90 heterocomplexes are immunoadsorbed from cytosol with a monoclonal antibody against the GR, complexes with each of these TPR domain proteins are seen. These large receptor heterocomplexes also contain cytoplasmic dynein, and immunoadsorption of FKBP52, CyP40 or PP5 from cytosol yields co-immunoadsorption of dynein. The large GR–hsp90–immunophilin–dynein complexes shown in Figure 4 can be formed by incubating immunopurified aporeceptor with reticulocyte lysate, and the presence of dynein in the complexes is selectively eliminated by competition with a purified PPIase domain fragment, showing that it is the PPIase domain that links the complex to the motor [24]. Overexpression of the PPIase domain fragment in cells eliminates the rapid movement of the GR along microtubular tracts [24] by disconnecting the immunophilin component of the receptor heterocomplex from the motor protein [23].

Thus the chaperone machinery interacts dynamically with the liganded GR to form GR–hsp90 heterocomplexes containing one of several TPR domain immunophilins that, in turn, link the receptor to the dynein motor responsible for retrograde movement to the nucleus. Because of the dynamic nature of complex assembly and disassembly, the receptors should be repeatedly associated with and released from the motor protein, yielding a saltatory pattern of movement, like that seen for vesicles along microtubular tracts.

Receptor cycling within the nucleus

Although hsp90 is located predominantly in the cytoplasm, it is clear that both hsp90 and hsp70 move into and out of the nucleus [1], and that the chaperone machinery functions within the nucleus [25]. It has been known for many years that the GR does not have DNA-binding activity when it is in heterocomplex with hsp90 and that the chaperone machinery in reticulocyte lysate acts on DNA-bound, hormone-free GR to convert the receptor into the non-DNA-binding state and restore steroid-binding activity [26]. This led to the proposal that hsp90 heterocomplex assembly is involved in the termination of transcriptional action when steroid dissociates from DNA-bound receptors in intact cells [26].

After the liganded GR has translocated to its sites of action in the nucleus and associated with the appropriate co-activators and response elements to activate transcription, the receptor cycles out of the transcription complexes as the hormone dissociates. It has been shown in permeabilized cells depleted of the cytoplasmic chaperone machinery that nuclear GR can recycle to the hormone-binding state, rebind hormone and target to chromatin-binding sites without ever exiting from the nucleus [27]. During hormone withdrawal, both release of the GR from chromatin and generation of high-affinity steroid-binding activity in the nucleus are inhibited by geldanamycin [27]. This provides quite clear evidence that the chaperone machinery is active within the nucleus and that it may facilitate the termination of transcriptional activation as steroid dissociates from the receptor.

Consistent with this notion, p23 has been shown to disrupt GR-mediated transcriptional activation both *in vitro* and *in vivo* [28]. In this case, it was not established that p23 was acting through its ability to bind to and stabilize the ATP-dependent conformation of GR-bound hsp90 as shown in Figures 2 and 3. However, it is a reasonable proposal that p23-promoted disruption of transcriptional regulatory complexes is mediated by the actions of the hsp90-/hsp70-based chaperone machinery on the hormone-free GR. It has been known for some time that hormone withdrawal is accompanied by simultaneous termination of transcriptional activation. Thus it has been thought that as the level of glucocorticoid decreases, the hormone is released from the receptor. However, as we have seen, the situation is far more complex because glucocorticoid binding to the GR at physiological temperature does not reflect an equilibrium binding state. When steroid dissociates, the GR must undergo assembly into a complex with hsp90 before it can bind steroid again and new transcriptional regulatory complexes can form. This chaperone-machinery-dependent assembly/disassembly cycle is probably occurring continually with transcriptional regulatory complexes in the nucleus, and when hormone levels decline, there is no free steroid to bind to newly assembled GR–hsp90 heterocomplexes and continue the cycle. Thus, the assembly/disassembly cycle allows for fine-tuned regulation of transcription according to changes in hormone level.

Receptor turnover

Treatment of cells with an hsp90 inhibitor, such as geldanamycin or radicicol, increases the degradation rate of many (probably all) client proteins that form persistent heterocomplexes with hsp90 [2,29]. When hsp90 function is inhibited, the GR and the other client proteins are degraded by the ubiquitin–proteasome pathway of proteolysis [30] as shown in Figure 5. The protein to be degraded becomes bound by a member of a family of ubiquitin ligases (E3), which brings the target protein and ubiquitin-charged E2 together into a complex for ubiquitin conjugation. Repetition of the process results in the formation of a polyubiquitin chain, and the polyubiquitylated protein is then recognized and degraded to small peptides by the 26 S proteasome complex, with the ubiquitin being recycled [30]. The GR and other steroid receptors are stabilized to ubiquitin–proteasomal degradation by being in complex with hsp90.

One E3 ligase for the GR is a 35 kDa cytoplasmic protein called CHIP (C-terminus of Hip) [31]. CHIP binds via a TPR domain to the C-terminus of both hsp70 and hsp90, and it possesses a C-terminal U-box that interacts with the ubiquitin-conjugating enzyme E2. CHIP and Hop bind to the same TPR acceptor sites on hsp70 and hsp90, and it has been suggested that competition between CHIP and Hop may modulate quality-control decisions between ubiquitin-dependent degradation and proper folding by the chaperone machinery [32]. However, it may be more productive to consider the triage decision from the viewpoint of the cleft-opening model.

As mentioned previously, the region where hydrophobic clefts emerge to the protein surface may be the general topological feature that is recognized by hsp70. If the cleft is such that it can be primed by hsp70, as is the unliganded GR in Figure 3, then the primed GR–hsp70 complex may preferentially bind Hop and hsp90 and proceed to GR–hsp90 heterocomplex assembly (Figure 6A). However, if the cumulative effect of oxidative and other damage to the receptor is sufficient, cleft opening will occur spontaneously as hydrophobic amino acids are exposed during the early stages of denaturation. Instead of encountering a closed cleft that can be primed for hsp90 assembly, hsp70 encounters a cleft opening with significant hydrophobic character where a primed complex cannot be achieved, and the hsp70 bound to the emerging hydrophobic cleft interior may prefer to bind CHIP and proceed towards degradation (Figure 6B). Thus hsp70

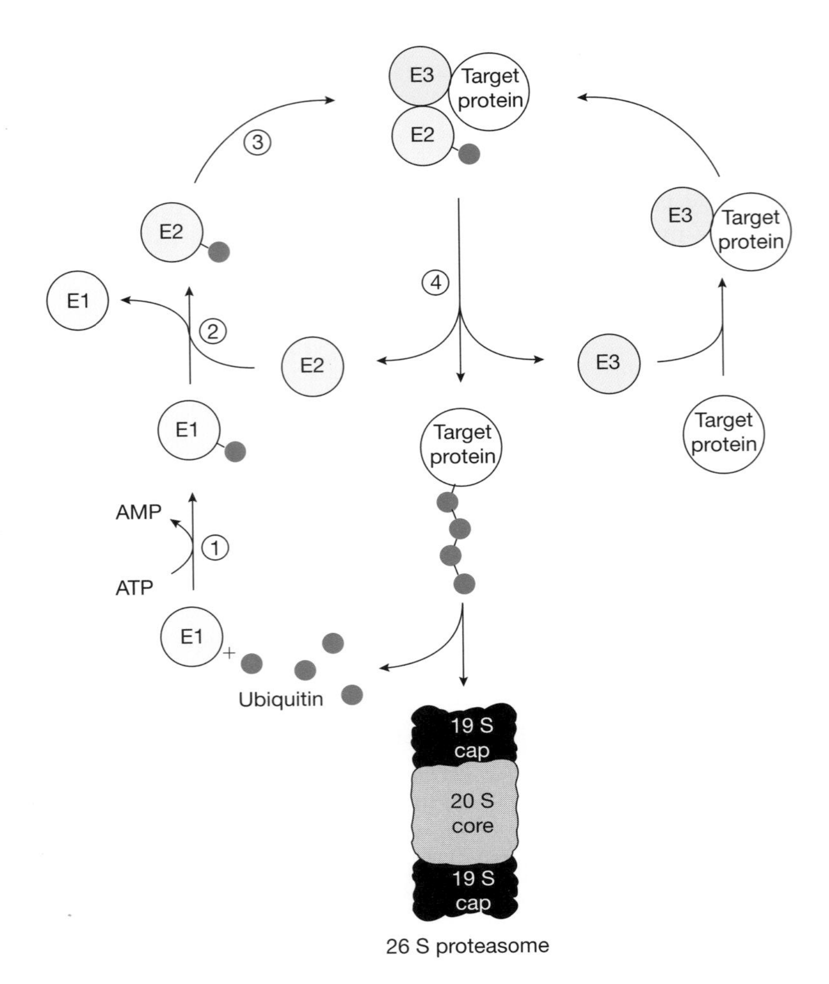

Figure 5. The ubiquitin–proteasome pathway of proteolysis
In an ATP-dependent reaction, the ubiquitin-activating enzyme E1 forms a high-energy thiolester bond with ubiquitin (1). The activated ubiquitin is then transferred in thiolester linkage to a member of the family of ubiquitin-conjugating enzymes, E2 (2). The protein targeted for degradation is bound by a member of the family of ubiquitin ligases, E3, which brings the target protein and ubiquitin-charged E2 together to form a complex (3). The ubiquitin now attaches to the target protein and further ubiquitins are conjugated to form a polyubiquitin chain (4). The polyubiquitylated substrates are recognized and degraded by the 26 S proteasome, which is composed of a 20 S core with proteolytic sites and two 19 S caps that recognize ubiquitin and contain ATPases that unfold proteins. The ubiquitin units themselves are recycled.

would make the triage decision in binding to the native (closed cleft) versus nascently denaturing (opening cleft) forms of the protein. We know that hsp70 conformation makes a difference, because in the absence of substrate, the ADP-bound conformation binds with higher affinity to Hop than the ATP-bound conformation [11]. How this preference is affected by the binding of substrate

(A)

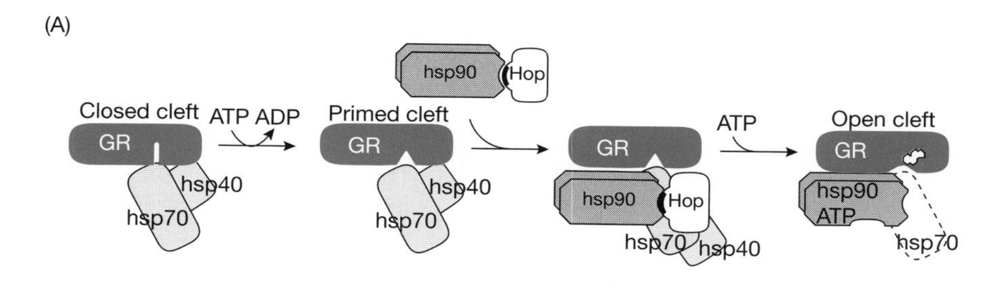

(B)

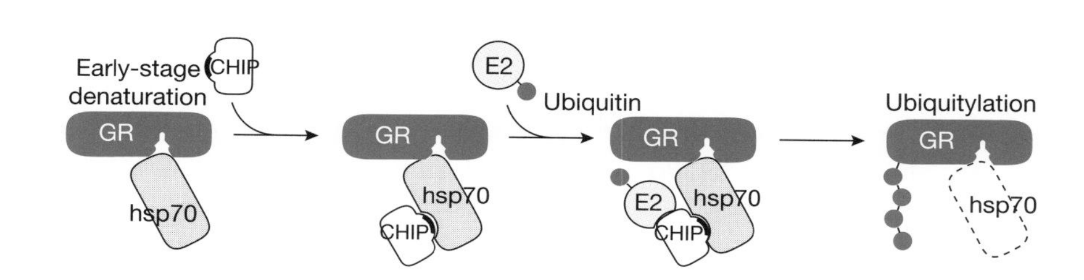

Figure 6. The triage decision between GR–hsp90 heterocomplex assembly and GR ubiquitylation and degradation
Hsp70 is shown here functioning as a sensor for protein triage decisions. In (**A**) hsp70 binds to the native state of the GR with a closed cleft to initiate GR–hsp90 heterocomplex assembly by forming a primed GR–hsp70 complex. In (**B**) hsp70 binds to a GR in the early stages of unfolding, and CHIP is bound, which, in turn, recruits ubiquitin-charged E2, leading to ubiquitylation and, ultimately, degradation by proteasome. As discussed in the text, these pathways are not exclusive; both hsp70 and CHIP are recovered in some native GR–hsp90 heterocomplexes.

protein is unknown, but it shows that the nucleotide-binding conformation of hsp70 affects the binding properties of its TPR acceptor site.

It is clear that pathways A and B in Figure 6 are not exclusive. After Hop leaves the TPR-binding site on GR-bound hsp70, the hsp70 can bind CHIP. Thus CHIP has been recovered with GR–hsp90 heterocomplexes, and it has been shown that both CHIP and hsp70 undergo retrograde movement with the GR that is hsp90-dependent. Although CHIP can be in the complex when hsp90 is bound to the GR, it apparently cannot bind ubiquitin-charged E2 and proceed with ubiquitylation. However, inhibition of hsp90 binding with geldanamycin or radicicol permits ubiquitylation and an increased rate of GR turnover.

Although CHIP can bind the TPR acceptor site in hsp90, it does not seem reasonable that its binding to GR-bound hsp90 signals receptor degradation. In all of the cases examined so far, formation of heterocomplexes with hsp90 stabilizes client proteins and inhibition of heterocomplex assembly by geldanamycin and radicicol promote degradation by the ubiquitin–proteasome pathway [2,29]. By extending the cleft-opening model of chaperone machinery action to other hsp90 client proteins, it is possible to envision how hsp70 may function as the sensor for these protein triage decisions.

Conclusion

We have seen here how hsp90 is important for regulating proper folding, trafficking, nuclear cycling and turnover of the GR. The study of steroid receptor–hsp90 heterocomplexes has led to the discovery of the hsp90-/hsp70-based chaperone machinery and to the hsp90-binding immunophilins. The chaperone machinery is present in all eukaryotic cells. Even lysates from plants can form GR–hsp90 heterocomplexes with high-affinity steroid-binding activity that bind plant TPR domain immunophilins that, in turn, bind cytoplasmic dynein via their PPIase domains [33]. The hsp90 and hsp70 components of the machinery are functionally conserved, in that the plant chaperones can replace their homologues when mixed with purified animal chaperone proteins, with full activity being retained. When nucleated cells evolved, the genes for hsp90 and hsp70 duplicated and hsp90 became an essential protein, like hsp70 [2]. It was probably at this time that hsp90 and hsp70 began to work together as a machinery performing fundamental cellular functions through their ability to interact with native proteins by recognizing regions where hydrophobic clefts merge with the protein surface. It is not, therefore, surprising that new client proteins are identified each month, and it is likely that most signal transduction pathways will be regulated in some way by the chaperone machinery.

Summary

- *Like many other proteins involved in signal transduction, the steroid receptors are regulated by hsp90.*
- *Heterocomplexes with hsp90 are formed by a multiprotein chaperone machinery in which hsp70 and hsp90 are the essential components and Hop, hsp40 and p23 function as co-chaperones.*
- *The chaperone machinery recognizes the native GR in the region where the hydrophobic ligand-binding cleft merges with the protein surface, and in making the GR–hsp90 heterocomplex the cleft is opened to access by a steroid.*
- *Receptor-bound hsp90 binds TPR domain immunophilins, such as FKBP52, CyP40 and PP5, which in turn connect the receptor to cytoplasmic dynein, the motor protein for retrograde trafficking to the nucleus.*
- *As steroid dissociates from the GR in transcriptional regulatory complexes, the chaperone machinery appears to detach the receptor, terminating the round of transcriptional activation.*

- *Formation of heterocomplexes with hsp90 prevents the ubiquitin ligase CHIP from promoting polyubiquitylation and receptor degradation via the proteasome.*

The work of the authors that led to some of the concepts described herein was supported by National Institutes of Heath grants CA28010 from the National Cancer Institute and DK31573 from the National Institute of Diabetes and Digestive and Kidney Diseases.

References

1. Pratt, W.B. & Toft, D.O. (1997) Steroid receptor interactions with heat shock protein and immunophilin chaperones. *Endocr. Rev.* **18**, 306–360
2. Pratt, W.B. & Toft, D.O. (2003) Regulation of signaling protein function and trafficking by the hsp90/hsp70-based chaperone machinery. *Exp. Biol. Med.* **228**, 111–133
3. Kaul, S., Murphy, P.J.M., Chen, J., Brown, L., Pratt, W.B. & Simons, S.S. (2002) Mutations at positions 547–553 of rat glucocorticoid receptors reveal that hsp90 binding requires the presence, but not defined composition, of a seven-amino acid sequence at the amino terminus of the li binding domain. *J. Biol. Chem.* **277**, 36223–36232
4. Kanelakis, K.C., Shewach, D.S. & Pratt, W.B. (2002) Nucleotide binding states of hsp70 and hsp90 during sequential steps in the process of glucocorticoid receptor•hsp90 heterocomplex assembly. *J. Biol. Chem.* **277**, 33698–33703
5. Mackem, S., Baumann, C.T. & Hager, G.L. (2001) A glucocorticoid/retinoic acid receptor chimera that displays cytoplasmic/nuclear translocation in response to retinoic acid. *J. Biol. Chem.* **276**, 45501–45504
6. Picard, D. (1993) Steroid-binding domains for regulating the functions of heterologous proteins *in cis*. *Trends Cell Biol.* **3**, 278–280
7. Billecke, S.S., Bender, A.T., Kanelakis, K.C., Murphy, P.J.M., Lowe, E.R., Kamada, Y., Pratt, W.B. & Osawa, Y. (2002) Hsp90 is required for heme binding and activation of apo-neuronal nitric-oxide synthase. *J. Biol. Chem.* **277**, 20504–20509
8. Hartl, F.U. & Hayer-Hartl, M. (2002) Molecular chaperones in the cytosol: from nascent chain to folded protein. *Science* **295**, 1852–1858
9. Picard, D. (2002) Heat-shock protein 90, a chaperone for folding and regulation. *Cell. Mol. Life Sci.* **59**, 1640–1648
10. Morishima, Y., Kanelakis, K.C., Silverstein, A.M., Dittmar, K.D., Estrada, L. & Pratt, W.B. (2000) The hsp organizer protein Hop enhances the rate of but is not essential for glucocorticoid receptor folding by the multiprotein hsp90-based chaperone system. *J. Biol. Chem.* **275**, 6894–6900
11. Hernandez, M.P., Sullivan, W.P. & Toft, D.O. (2002) The assembly and intermolecular properties of the hsp70–Hop–hsp90 molecular chaperone complex. *J. Biol. Chem.* **277**, 38294–38304
12. Murphy, P.J.M., Kanelakis, K.C., Galigniana, M.D., Morishima, Y. & Pratt, W.B. (2001) Stoichiometry, abundance, and functional significance of the hsp90/hsp70-based multiprotein chaperone machinery in reticulocyte lysate. *J. Biol. Chem.* **276**, 30092–30098
13. Morishima, Y., Murphy, P.J.M., Li, D.P., Sanchez, E.R. & Pratt, W.B. (2000) Stepwise assembly of glucocorticoid receptor–hsp90 heterocomplex resolves two sequential ATP-dependent events involving first hsp70 and then hsp90 in opening of the steroid binding pocket. *J. Biol. Chem.* **275**, 18054–18060
14. Hernandez, M.P., Chadli, A. & Toft, D.O. (2002) HSP40 binding is the first step in the HSP90 chaperoning pathway for the progesterone receptor. *J. Biol. Chem.* **277**, 11873–11881

15. Sullivan, W. P., Owen, B.A.L. & Toft, D.O. (2002) The influence of ATP and p23 on the conformation of hsp90. *J. Biol. Chem.* **277**, 45942–45948
16. Morishima, Y., Kanelakis, K.C., Murphy, P.J.M., Shewach, D.S. & Pratt, W.B. (2001) Evidence for iterative ratcheting of receptor-bound hsp70 between its ATP and ADP conformations during assembly of glucocorticoid receptor–hsp90 heterocomplexes. *Biochemistry* **40**, 1109–1116
17. Panaretou, B., Siligardi, G., Meyer, P., Maloney, A., Sullivan, J.K., Singh, S., Millson, S.H., Clarke, P.A., Noaby-Hansen, S., Stein, R. et al. (2002) Activation of the ATPase activity of hsp90 by the stress-regulated cochaperone Aha1. *Mol. Cell* **10**, 1307–1318
18. Schneikert, J., Hubner, S., Langer, G., Petri, T., Jaattela, M., Reed, J. & Cato, A.C.B. (2000) Hsp70–RAP46 interaction in downregulation of DNA binding by the glucocorticoid receptor. *EMBO J.* **19**, 6508–6516
19. Schmidt, U., Wochnik, G.M., Rosenhagen, M.C., Young, J.C., Hartl, F.U., Holsboer, F. & Rein, T. (2003) Essential role of the unusual DNA binding motif of BAG-1 for inhibition of the glucocorticoid receptor. *J. Biol. Chem.* **278**, 4926–4931
20. DeFranco, D.B., Madan, A.P., Tang, Y., Chandran, U.R., Xiao, N. & Yang, J. (1995) Nuclear cytoplasmic shuttling of steroid receptors. *Vitam. Horm.* **51**, 315–338
21. Pratt, W.B., Silverstein, A.M. & Galigniana, M.D. (1999) A model for the cytoplasmic trafficking of signalling proteins involving the hsp90-binding immunophilins and $p50^{cdc37}$. *Cell Signal.* **11**, 839–851
22. Hirokawa, N. (1998) Kinesin and dynein superfamily proteins and the mechanism of organelle transport. *Science* **279**, 519–526
23. Galigniana, M.D., Harrell, J.M., Murphy, P.J.M., Chinkers, M., Radanyi, C., Renoir, J.M., Zhang, M. & Pratt, W.B. (2002) Binding of hsp90-associated immunophilins to cytoplasmic dynein: direct binding and *in vivo* evidence that the peptidylprolyl isomerase domain is a dynein interaction domain. *Biochemistry* **41**, 13602–13610
24. Galigniana, M.D., Radanyi, C., Renoir, J.M., Housley, P.R. & Pratt, W.B. (2001) Evidence that the peptidylprolyl isomerase domain of the hsp90-binding immunophilin FKBP52 is involved in both dynein interaction and glucocorticoid receptor movement to the nucleus. *J. Biol. Chem.* **276**, 14884–14889
25. DeFranco, D.B. (2000) Role of molecular chaperones in subnuclear trafficking of glucocorticoid receptors. *Kidney Int.* **57**, 1241–1249
26. Scherrer, L.C., Dalman, F.C., Massa, E., Meshinchi, S. & Pratt, W.B. (1990) Structural and functional reconstitution of the glucocorticoid receptor–hsp90 complex. *J. Biol. Chem.* **265**, 21397–21400
27. Liu, J. & DeFranco, D.B. (1999) Chromatin recycling of glucocorticoid receptors: implications for multiple roles of heat shock protein 90. *Mol. Endocrinol.* **13**, 355–365
28. Freeman, B.C. & Yamamoto, K.R. (2002) Disassembly of transcriptional regulatory complexes by molecular chaperones. *Science* **296**, 2232–2235
29. Neckers, L., Schulte, T.W. & Mimnaugh, E. (1999) Geldanamycin as a potential anti-cancer agent: its molecular target and biochemical activity. *Invest. New Drugs* **17**, 361–373
30. Doherty, F.J., Dawson, S. & Mayer, R.J. (2002) The ubiquitin–proteasome pathway of intracellular proteolysis. In *Essays in Biochemistry*, vol. 38 (Hooper, N.M., ed.), pp. 51–63, Portland Press, London
31. Connell, P., Ballinger, C.A., Jiang, J., Wu, Y., Thompson, L.J., Hohfeld, J & Patterson, C. (2001) The co-chaperone CHIP regulates protein triage decisions mediated by heat shock proteins. *Nat. Cell Biol.* **3**, 93–96
32. Hohfeld, J., Cyr, D.M. & Patterson, C. (2001) From cradle to the grave: molecular chaperones that may choose between folding and degradation. *EMBO Rep.* **2**, 885–890
33. Harrell, J.M., Kurek, I., Breiman, A., Radanyi, C., Renoir, J.M., Pratt, W.B. and Galigniana, M.D. (2002) All of the protein interactions that link steroid receptor–hsp90–immunophilin heterocomplexes to cytoplasmic dynein are common to plant and animal cells. *Biochemistry* **41**, 5581–5587

5

DNA recognition by nuclear receptors

Frank Claessens*[1] and Daniel T. Gewirth†

**Department of Molecular Cell Biology, Faculty of Medicine, Campus GHB O/N, Herestraat 49, 3000 Leuven, Belgium, and †Department of Biochemistry, Duke University Medical Center, Durham, NC, U.S.A.*

Abstract

The nuclear receptors constitute a large family of ligand-inducible transcription factors. The control of many genetic pathways requires the assembly of these nuclear receptors in defined transcription-activating complexes within control regions of ligand-responsive genes. An essential step is the interaction of the receptors with specific DNA sequences, called hormone-response elements (HREs). These response elements position the receptors, and the complexes recruited by them, close to the genes of which transcription is affected. HREs are bipartite elements that are composed of two hexameric core half-site motifs. The identity of the response elements resides in three features: the nucleotide sequence of the two core motif half-sites, the number of base pairs separating them and the relative orientation of the motifs. The DNA-binding domains of nuclear receptors consist of two zinc-nucleated modules and a C-terminal extension. Residues in the first module determine the specificity of the DNA recognition, while residues in the second module are involved in dimerization. Indeed, nuclear receptors bind to their HREs as either homodimers or heterodimers. Depending on the type of receptor, the C-terminal extension plays a role in sequence recognition, dimerization, or both. The DNA-binding domain is furthermore involved in several other

[1]*To whom correspondence should be addressed (e-mail Frank.claessens@med.kuleuven.ac.be).*

functions including nuclear localization, and interaction with transcription factors and co-activators. It is also the target of post-translational modifications. The DNA-binding domain therefore plays a central role, not only in the correct binding of the receptors to the target genes, but also in the control of other steps of the action mechanism of nuclear receptors.

Introduction

Transcription factors occupy the functional nexus between the stored information of the DNA genome and the extraction and expression of that information in the life activity of the cell. The superfamily of nuclear hormone receptors, of which there are at least 150 known members, has evolved to combine the disparate functions of signal responsiveness, DNA binding and transcriptional activation into one protein composed of functionally separate domains. DNA-binding domains (DBDs), the subject of this review, must overcome the challenge of finding a small cognate response element amid an 8×10^9 bp sea of cellular DNA. This would ordinarily be a formidable-enough challenge, but it is made even more difficult by the fact that the set of hormone-response elements (HREs) is itself quite limited in both structure and sequence. Moreover, a distinguishing feature of nuclear receptor DBDs is that they are highly conserved, even among the most distant relatives within the family [1–4]. A long-standing question is, therefore, how site selection is achieved, given that nuclear receptors employ a highly conserved DBD and target one of only two types of hexameric half-site sequence. In this chapter, we will discuss in more detail the current status and specifics of the HREs and of the DNA recognition by nuclear receptors.

Figure 1. Consensus binding sequences for the different classes of nuclear receptors
(**A**) The nuclear receptor superfamily has been divided into four classes depending on the type of interaction with the HREs. The DNA-binding sites for nuclear receptors exist as one or two copies of hexamer sequences, which resemble either 5′-AGAACA-3′ for the class I receptors or 5′-AGGTCA-3′ for the other receptors. Steroid receptors only bind inverted repeats with a three-nucleotide spacer (IR3), except for the androgen receptor (AR) which also binds DR3 [19–22]. DRs of the 5′-AGGTCA-3′ core are recognized by heterodimers of 9-*cis* retinoic acid receptor (RXR) with other nuclear receptors, except for those receptors indicated with an asterisk which can homodimerize on the cognate response element. The number of spacing nucleotides restricts the species of receptor dimers that can activate through these HREs; this was called the 1 to 5 rule [4]. Monomer-binding sites for class IV receptors are extended along their 5′ side as indicated by the nucleotides given in lower case. It should be noted that this is not a complete listing. CAR, constitutively active receptor; ERR, ER-related receptor; GR, glucocorticoid receptor; MR, mineralocorticoid receptor; PR, progesterone receptor; ER, oestrogen receptor; PPAR, peroxisome-proliferator-activated receptor; COUP, chicken ovalbumin upstream promotor receptor; RAR, all-*trans* retinoic acid receptor; VDR, vitamin D receptor; T3R, thyroid receptor; PXR, pregnane X receptor; LXR, liver X receptor; NGFI-B, nerve growth factor-inducible factor I-B. (**B**) Illustration of the hypothesis of the conformational changes taking place during DNA-dependent dimerization of the DBDs. As an example, two DBDs of a class I receptor is shown (light blue boxes), one of which first binds to the most conserved hexamer with high affinity. The subsequent conformational change in the DBD–DNA complex (step 1, dark blue) results in a co-operative recruitment of the dimeric partner (step 2) to a downstream hexamer. The conformational change, however, does not allow co-operative binding to an upstream-situated hexamer [20,21]. ☞

The HREs

A comparison of consensus HRE sequences for the steroid receptors (Figure 1A) reveals that they typically recognize palindromic inverted repeats of two hexameric core sequences, separated by three nucleotides (IR3). Early experiments showed that the androgen receptor (AR), glucocorticoid receptor (GR), mineralocorticoid receptor and progesterone receptor (PR), also called the class I receptors, all recognize a 5′-AGAACA-3′ core, while the oestrogen receptors (ERs), also called the class II receptors, recognize a 5′-AGGTCA-3′ core [2,5]. The class I receptor core sequence has proven to be the exception in

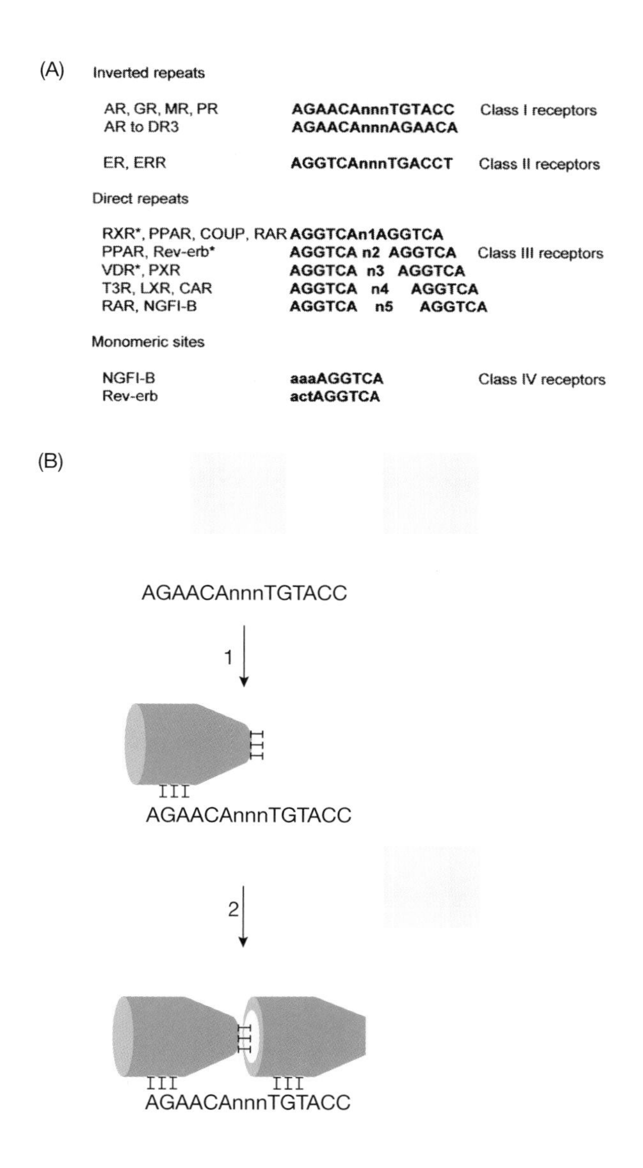

the nuclear receptor superfamily, as all other nuclear receptors bind preferably to the 5′-AGGTCA-3′ sequence (Figure 1A).

With only two consensus half-site sequences and over 150 different hormone receptors, additional diversity must be generated topologically, and this is achieved largely by varying the arrangement of the half-sites relative to one another. Inverted, everted and direct repeats (DRs) all restrict the dimeric receptor species that can bind. Among the best-characterized non-palindromic arrangements are the DRs of 5′-AGGTCA-3′, which are targets for the nuclear receptors that form heterodimers with RXR (9-*cis* retinoic acid receptor) [3,4]. The repeats vary in the length of the spacer that separates the two hexameric half-sites, and in this way, DRs 1, 2, 3, 4 and 5 are each recognized by specific receptor dimers. This was named the 1 to 5 rule (Figure 1A).

The spacing between the cores affects the position of the two bound nuclear receptor monomers since the addition of 1 bp results in a relative rotation of approx. 35° and a 3.4 Å (1Å = 0.1 nm) increase in their separation. This means that the dimerization surfaces of the nuclear receptors on different DR targets must be very different. This contrasts with the class I receptors, which all bind to IR3 repeats of the 5′-AGAACA-3′ core element. Since the spacing of these cores is always 3 bp, the dimerization interface for these receptors is likely to be unchanged from one steroid receptor to another (Figure 1A).

Although there are only two consensus core sequences for HREs, the two halves of the bipartite response element are not equal. Indeed, early analyses comparing all natural steroid-response elements revealed that one half-site is on average more like the consensus sequence while the other can diverge from this consensus considerably [2]. Together with the crystal structure of the GR DBD complexed with DNA [6], a mechanism of receptor binding was proposed wherein the consensus half-site is recognized with high affinity first. The binding of the first DBD results in a conformational change in the protein that supports the formation of the proper dimerization interface, thereby enabling a second monomer to bind co-operatively to the second, less conserved, half-site element (Figure 1B). Remarkably, the DBD dimerization interface was found to be of such strength that when challenged with an IR4 response element instead of the preferred IR3, the protein–protein interface was preserved, even at the expense of forcing the second DBD to bind to a non-cognate half-site. This suggested a model whereby the set of two half-sites that together allows the lowest energy DBD dimerization interface has the highest overall affinity, even though the half-site elements individually may not have the highest affinity. Thus the second half site, although it is far less conserved, can modulate the affinity of the receptor for the HRE. For example, while some mutations in the consensus half-site abolish binding completely, the same mutations in the second half-site only moderately increase or decrease the binding affinities of steroid receptors [7,8]. This model, that distortions of the DBD dimer interface are energetically more costly than sub-optimal DBD–half-site interactions, also explains why the consensus high-affinity binding site (5′-GGTACANNNTGTTCT-3′) for GR, PR and AR is not a perfect inverted repeat [7,9]. Support for this model also comes from several studies that have shown that the nucleotide identity in the spacer of steroid-response elements, which can affect the precise stereochemical relationship of the two half-site elements, is not neutral to DNA binding by the DBDs and to transactivation. Indeed, specific preferences in PCR-mediated selections of receptor-binding elements as well as in natural elements correlate with increased steroid responsiveness in transfection experiments (e.g. [7,10,11]).

The DBDs

When the cDNAs encoding the ER and GR were first compared, the DBDs were identified on the basis of their sequence conservation and high proportion of basic residues. The DBD is centrally located within the steroid receptors and contains eight cysteine residues whose conserved spacing immediately suggested that this domain would bind zinc ions. Indeed, the minimum DBD consists of approx. 63 amino acids and contains two zinc-binding modules, with each zinc ion in a tetrahedral arrangement co-ordinated by four cysteine residues [1]. The DBD contains several conserved hydrophobic and aromatic residues that form a compact, hydrophobic core [9]. Together with the two Cys_4-Zn clusters, the hydrophobic core stabilizes the globular structure of the DBD. The combination of a hydrophobic core and the co-ordinated zinc ion distinguishes the nuclear hormone receptors from the classical zinc finger structures.

The two zinc-nucleated modules are encoded by separate exons and have different functions. Module-swapping experiments between different steroid receptors showed that the first zinc-binding module is involved in sequence specificity, i.e. the discrimination between 5′-AGAACA-3′ and 5′-AGGTCA-3′ (e.g. [12]). Within this first module, mutation analysis revealed a proximal box or P-box, which contains the residues necessary for sequence discrimination (Figure 2A). In contrast, the second module does not confer sequence specificity but instead contains several residues which form the distal box or D-box segment involved in a DNA-dependent dimerization of the class I and class II DBDs (Figure 2B) [9].

Discrimination of the core hexamers

Luisi et al. [6] solved the structure of the GR DBD in complex with DNA. When bound to DNA, the DBD makes extensive contacts with the sugar-phosphate backbone of both strands such that the recognition helix rests in the major groove roughly perpendicular to the helical axis. In this orientation, the P-box residues are in position to make either direct or water-mediated contacts with the major groove faces of the base pairs. For the GR, and by inference all class I receptors, this involves contacts between the residues KVXXXR and

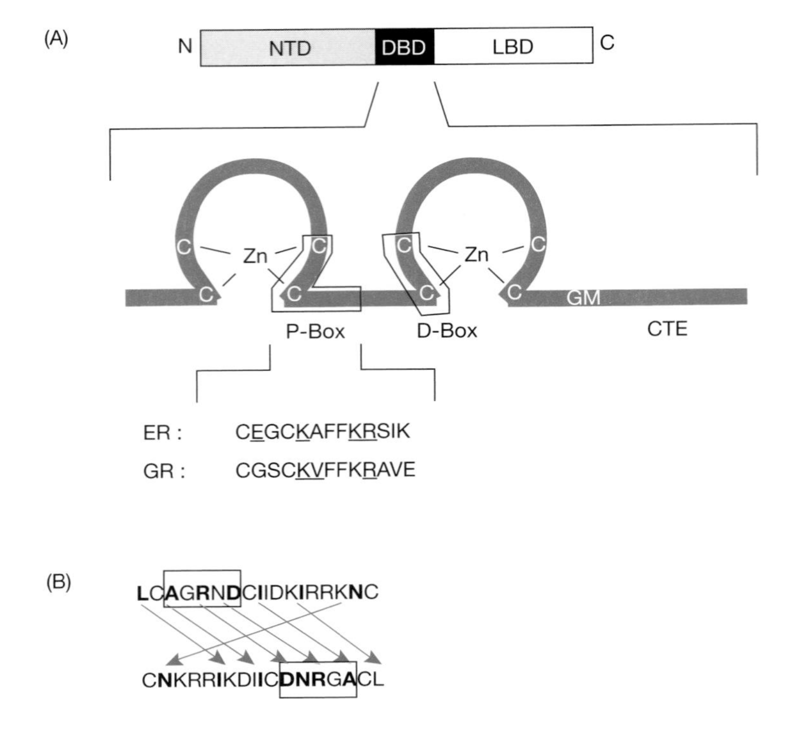

Figure 2. (A) Two types of P-box sequence and (B) GR dimerization interface
(**A**) The DBDs of nuclear receptors are located in the centre of the protein. Two zinc-coordinated modules are drawn as zinc fingers, and the locations of the P- and D-boxes are indicated. The DNA-binding specificity by nuclear receptors is dictated by a DNA-recognition helix which inserts in the major groove of the DNA helix. For the class I receptors (represented by the GR) on the one hand, and the class II, III and IV receptors (represented by the ER) on the other hand, the DNA-recognition helices differ in sequence as indicated. The residues making contact with the DNA or contributing to the sequence selectivity are underlined (based on [6,13]). NTD, N-terminal domain; LBD, ligand-binding domain. (**B**) The peptide fragments that encompass the D-box (in boxes) are given and the different contacts between the residues (in bold) of two dimerizing GR DBDs are indicated by arrows (based on [6]). The lower D-box sequence is reversed to represent the conformational alignment in the monomers. Dimerization of the ER DBDs involves very similar complementarity of shape of the corresponding region [13].

the three bases underlined in the upper (5′-AGAACA-3′) and lower strand (5′-TGTTCT-3′) of the half-site (Figure 3A). The ER has a different set of P-box amino acids on its recognition helix, and in this case the structure of the ER DBD–DNA complex showed that contacts are made between residues EXXKXXXKR and the bases underlined in the upper (5′-AGGTCA-3′) and lower strand (5′-TGACCT-3′; Figure 3B) [13].

Discrimination between the 5′-AGAACA-3′ class I element and the 5′-AGGTCA-3′ ER-/nuclear receptor-response element by the DBDs is mainly due to differences in the helical structure of the response elements, which leads to the inclusion of additional water molecules between the DNA and the protein when the DBD is bound to the incorrect core element. This destabilizes

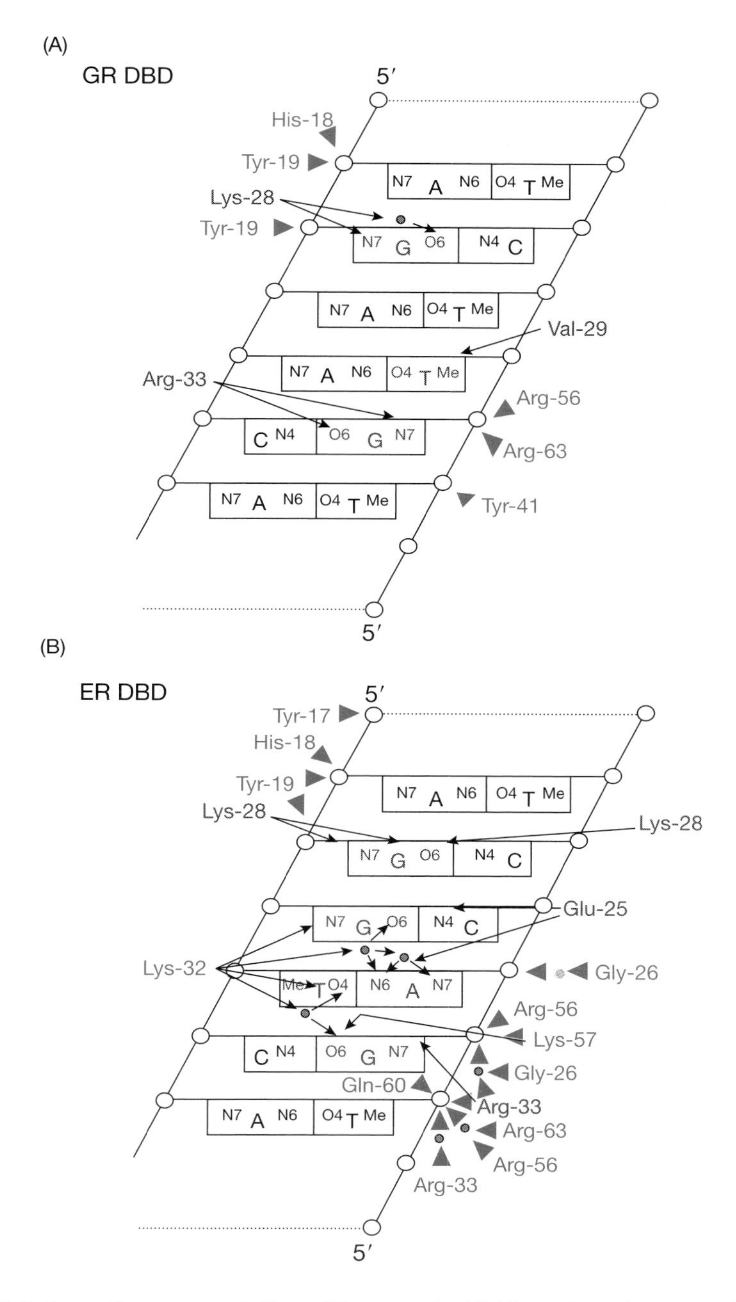

Figure 3. Schematic representation of the protein–DNA contacts between the DNA-recognition helices of the DBDs of GR (A) versus ER (B) to 5′-AGAACA-3′ and 5′-AGGTCA-3′ respectively

The major groove of the hexamer is projected and the interactions between receptor residues and the DNA are indicated by arrows. Phosphate-contacting residues are given in blue, base-contacting residues are given in grey. Small filled circles represent water molecules. Several amino acid side chains can make multiple contacts with DNA (e.g. Arg-33 in the ER).

the interaction due to the increased entropic burden [14]. An additional increase in the discriminative power of the DBDs has been suggested to involve negative interactions of P-box residues, which prevent binding to non-cognate response elements (reviewed in [9]).

Dimerization controls specificity

An important feature of many nuclear receptor DBDs is their ability to dimerize in the presence of their cognate DNA sites. In the absence of DNA, protein–protein interactions between DBDs do not occur because the potential dimerization surfaces are small. DNA-dependent dimerization thus adds to the specificity of sequence recognition on the one hand, and to the diversity of sequence elements which can be recognized on the other. For the steroid receptors, the separation of the hexamer cores by three nucleotides is a prerequisite for high-affinity binding (e.g. [7,10]). By forming a co-operative dimer, the receptors measure both the spacing and the helical repeat of the response element, thus greatly increasing the specificity of the interaction. For the class III receptors, the possibility of heterodimerization with RXR expands the diversity of the recognition sequences (Figure 1A) [15].

So far, we have mainly discussed the binding of nuclear receptor DBDs to IRs. However, many nuclear receptors recognize DRs either as homodimers or, more commonly, as heterodimers with RXR. The mechanism of spacer discrimination on DR response elements was revealed by the structures of several nuclear receptor DBDs bound to cognate DR targets [15–17]. This series of structures highlighted the role of the C-terminal extension (CTE) to the core DBD. The CTEs of nuclear receptor DBDs bound to DR-response elements are arranged such that the CTE of the downstream partner is positioned to make specifying contacts with the upstream protein. By contrast, the CTEs of the steroid receptors (GR, ER) bound to palindromic targets are on opposing faces of the DBD dimer and are not positioned to make cross-dimer interactions [6,13].

CTEs vary widely in sequence between receptors, and this is reflected in considerable structural variation. In the case of vitamin D receptor (VDR) and thyroid hormone receptor (TR) [15–17], the CTE forms a 24-residue α-helix that extends nearly to the start of the ligand-binding domain. The seven residues of the CTE between the DBD core and the α-helix make protein–protein contacts with the second zinc-coordinating module of the RXR partner, thus forming part of the dimerization interface. For the TR–RXR heterodimer, the TR CTE is also involved in contacts with the minor groove of the spacer sequence, thus explaining the sequence preferences of this heterodimer in this part of the HRE. The long α-helix serves as a steric 'ruler' that restricts binding to correctly spaced core elements by making van der Waals clashes with the upstream partner on incorrectly spaced targets. The CTEs of some other nuclear receptors do not employ a rigid secondary structural element for spacer selectivity and dimerization. Instead, as in the case of orphan receptors of class IV (Figure 1A) that bind DNA as monomers, they use specific CTE–DNA contacts to enlarge their recognition sequence, enabling further discrimination on the basis of nucleotide identity immediately upstream of the core hexamer [15].

While the role of the CTE is clear for the class III and class IV receptors, its role in steroid receptor DNA binding is less well understood. The CTEs in the structures of the GR DBD and ER DBD were not ordered [6,13], even though deletion studies implicated the CTE in DNA binding, if not DNA recognition [18]. As discussed next, the AR might be an exception to the rules on DNA binding by the steroid receptors as inferred from GR and ER DBD structural data.

Exceptions to the rules

It has been known for some time that steroid receptors bind with high affinity to IR3s, and most steroid response elements fall into this category. Not surprisingly, DNA targets containing IR3s of the 5′-AGAACA-3′ core confer simultaneous androgen, glucocorticoid, mineralocorticoid and progesterone responsiveness to heterologous promoters [5,7–10]. However, some elements characterized in androgen target genes are exclusively activated by the AR, a puzzling result given that all formerly known androgen-response elements were IR3s, and that these IR3 elements were recognized equally well by all steroid receptors [7,19,20]. Swapping experiments were subsequently able to show that the second zinc-binding module and the CTE are involved in the binding to these AR-specific elements [21]. Because these parts of the DBD are not well positioned to make specifying contacts with the DNA bases, this suggested that an alternative dimerization mode, rather than alternative sequence specificity, was the basis for this AR-selective phenomenon. Indeed, upon careful examination of the selective androgen-response elements, it was noted that they might be DR3s of the 5′-AGAACA-3′ core element, rather than IRs [22]. The surprising implication of this analysis is that the AR DBD may have two dimerization surfaces, allowing it to bind to IR3s as a classical steroid receptor, or, alternatively, to DR3s in a mode that resembles a nuclear receptor, VDR (Figure 4) [17].

For the other steroid receptors, *in vitro* assays have demonstrated they do not bind to DR elements [8,11]. Steroid receptors have evolved from a predecessor in common with the other nuclear receptors. This predecessor probably bound DNA as a monomer and subsequently evolved into nuclear receptors, which gained the ability to bind dimerically to DRs. We hypothesize that while the other steroid receptors have lost their ability to bind DRs, the AR seems to have retained this characteristic. Therefore, although many other mechanisms are involved in the specificity of the *in vivo* responses of steroids,

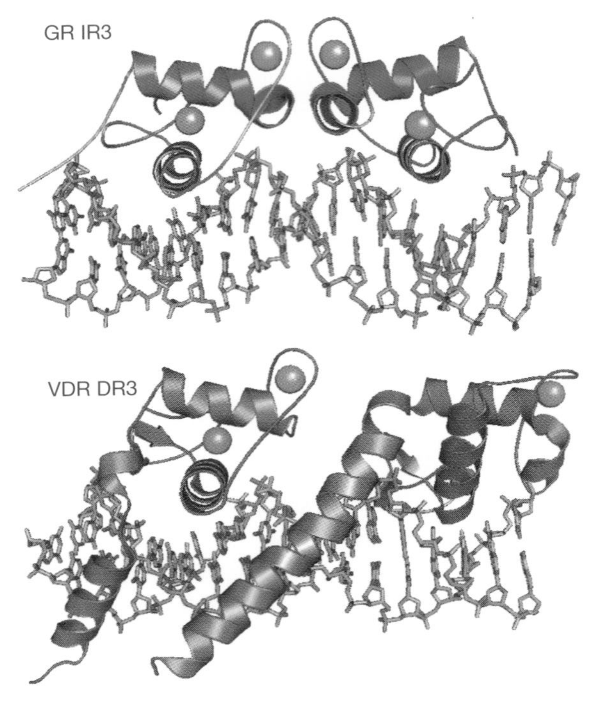

Figure 4. Structure of the GR DBD dimer (blue) binding to a three nucleotide spaced IR (top panel) clearly shows the head-to-head dimerization and the α-helix entering the major groove of the HRE [6]
The two zinc residues are indicated by purple spheres. The structure of the dimer of the VDR DBD (green) on a DR separated by three nucleotides is shown in the lower panel [16]. The long α-helical CTE of the right-hand monomer folding back and making protein–protein contacts with the left-hand monomer is clearly visible.

alternative DNA recognition by the AR explains part of the androgen-selective activation of target genes [20].

Allosteric effects of HREs

The sequence of the HRE not only serves as a docking site for receptor dimers. It can also dictate, probably via allosteric effects, whether the receptor will activate or repress transcription of a neighbouring promoter. Indeed, GR binding to sites in the prolactin and the proliferin genes result in a glucocorticoid-mediated repression, and single-base-pair mutations convert these negative response elements into positive ones [23]. AR binding to different androgen-response elements leads to alternative mechanisms not only in the dimerization of the DBDs, but also in subsequent steps in gene activation, e.g. co-activator recruitment by the activation functions 1 and 2 (AF1 and AF2) [24,25]. In general, allosteric effects of ligand and DNA on nuclear receptors seems to determine which co-activator complexes are recruited and more importantly what activities the recruited complexes will have [26].

Other functions of the DBD

Nuclear receptors do not occupy their response elements continuously, and the free DBD mediates other functions such as nuclear localization, nuclear export, interactions with chaperones and communication with other transcription factors (reviewed in [27]). Similarly, even DNA-bound DBDs can have additional functions. For example, the C-terminal half of the DBD and hinge region, by virtue of not making extensive contacts with the DNA, are able to serve as the recruitment site for several co-activators [28]. ARIP (AR-interacting protein), as a putative co-activator, as well as enzymes involved in sumoylation of the receptor, have been isolated in double-hybrid screening experiments with AR fragments. Specific lysine residues within the CTE of the ER and AR have been shown to be substrates for acetylation by pCAF (p300/CBP-interacting factor) and Tip60 (Tat interactive protein of 60 kDa). This acetylation is clearly involved in a regulation of the activity of these receptors, but the exact molecular mechanisms remain elusive [29].

General conclusions

HREs are embedded in complex enhancers to which several transcription factors can bind. Nuclear receptors will therefore become part of larger protein complexes, sometimes called enhanceosomes, that affect chromatin structure and hence transcription of the underlying genes [30]. It will be an exciting challenge to find out how the different signals coming through the nuclear receptors via the binding of ligands and DNA sequences, through the post-translational modifications, and through the recruited co-activator or co-repressor complexes, all become integrated with those coming from other signal pathways at the sites of these enhanceosomes.

Note added in proof

A crystal structure of the AR DBD bound to a three-spacer direct repeat steroid response element has been determined recently [31]. Surprisingly, this structure shows that the AR DBD binds in a symmetrical head-to-head arrangement, like the GR and ER DBDs, instead of the expected head-to-tail arrangement. In this conformation, the AR DBD dimer interface is more stable than the equivalent interface in the GR DBD. This implies that selective HREs that appear to have alternative arrangements of their hexameric half sites may be further examples of the ability of these receptors to exploit the strength of their DBD dimerization interfaces to accommodate sub-optimal protein–half-site interactions. This is likely to be not only a mechanism of response element discrimination, but also an effective way of modulating transcription from different hormone responsive genes. Finally, it will be interesting to study the exact contribution of the C-terminal extension of the DBD, of which the structure remains unsolved, to DNA binding and transactivation by the AR.

Summary

- *Nuclear receptors are ligand-inducible transcription factors which recruit a series of transcription-modulating complexes to specific DNA sequences located near their target gene.*
- *The DNA elements are typically organized as direct or inverted repeats of the hexamers 5′-AGAACA-3′ or 5′-AGGTCA-3′, and the relative orientation and distance between the repeats is the main determinant for receptor specificity.*
- *The DBDs are organized as two zinc-coordinating modules, the first of which makes specific contacts with the hexamer sequences. The second zinc-coordinating module serves as a dimerization interface.*
- *Depending on the type of receptor, a short CTE of the DBD plays a crucial role in sequence recognition through direct DNA interactions, or through a dimerization. The latter fixes the relative positions of the DNA-recognition residues and hence dictates the optimal distance between the repeats in the HREs.*
- *The AR is the only steroid receptor that recognizes both IRs and DRs of the 5′-AGAACA-3′ hexamer, probably through two alternative dimerization interfaces.*
- *Besides DNA binding and dimerization, the DBDs have multiple functions in transcriptional control.*

We thank the Belgian Fund for Scientific Research, the Catholic University of Leuven (K.U. Leuven, Belgium) and the DOD Prostate Cancer Research Program (DAMD17-02-1-0082) for support of the work which led to some of the findings and concepts described in this chapter. We are grateful to the members of our laboratories for fruitful discussions and interactions.

References

1. Evans, R.M. (1988) The steroid and thyroid hormone receptor superfamily. *Nature (London)* **240**, 889–895
2. Beato, M. (1989) Gene regulation by steroid hormones. *Cell* **56**, 335–344
3. Rastinejad, F., Perlmann, P., Evans, R. & Sigler, P. (1995) Structural determinants of nuclear receptor assembly on DNA direct repeat. *Nature (London)* **375**, 203–211
4. Gronemeyer, H. & Moras, D. (1995) How to finger DNA. *Nature (London)* **375**, 190–191
5. Cato, A.C.B., Miksicek, R., Schutz, G., Arnemann, J. & Beato, M. (1986) The hormone regulatory element of mouse mammary tumour virus promotes progesterone induction. *EMBO J.* **5**, 2237–2240
6. Luisi, B.F., Xu, W.X., Otwinowski, Z., Freedman, L.P., Yamamoto, K.R. & Sigler, P.B. (1991) Crystallographic analysis of the interaction of the glucocorticoid receptor with DNA. *Nature (London)* **352**, 497–505

7. Ham, J., Thomson, A., Needham, M., Webb, P. & Parker, M. (1988) Characterization of response elements for androgens, glucocorticoids and progestins in mouse mammary tumour virus. *Nucleic Acids Res.* **16**, 5263–5276
8. Schoenmakers, E., Alen, P., Verrijdt, G., Peeters, B., Verhoeven, G., Rombauts, W. & Claessens, F. (1999) Differential DNA binding by the androgen and glucocorticoid receptors involves the second Zn-finger and a carboxy terminal extension of the DNA-binding domains. *Biochem. J.* **341**, 515–521
9. Zilliacus, J., Wright, A., Carlstedt-Duke, J. & Gustaffson, J.-Å. (1995) Structural dterminants of DNA-binding specificity by steroid receptors. *Mol. Endocrinol.* **9**, 389–400
10. Roche, P.J., Hoare, S.A. & Parker, M.G. (1992) A consensus DNA-binding site for the androgen receptor. *Mol. Endocrinol.* **6**, 2229–2235
11. Haelens, A., Verrijdt, G. Callewaert, L., Christiaens, V., Schauwaers, K., Peeters, B., Rombauts, W. & Claessens, F. (2002) DNA recognition by the androgen receptor: evidence for an alternative DNA-dependent dimerization, and an active role of sequences flanking the response element on transactivation. *Biochem. J.* **369**, 141–151
12. Green, S., Kumar, V., Theulaz, I., Wahli, W. & Chambon, P. (1988) The N-terminal DNA-binding zinc finger of the oestrogen and glucococrticoid receptors determines target gene specificity. *EMBO J.* **7**, 3037–3044
13. Schwabe, J.W.R., Chapman, L., Finch, J.T. & Rhodes, D. (1993) The crystal structure of the estrogen receptor DNA-binding domain bound to DNA: how receptors discriminate between their response elements. *Cell* **75**, 567–578
14. Gewirth, D.T. & Sigler, P.B. (1995) The basis for half-site specificity through a non-cognate steroid receptor-DNA complex. *Nat. Struct. Biol.* **2**, 386–394
15. Rastinejad, F. (2001) Retinoid X receptor and its partners in the nuclear receptor family. *Curr. Opin. Struct. Biol.* **11**, 33–38
16. Shaffer, P.L. & Gewirth, D.T. (2002) Structural basis of VDR-DNA interactions on direct repeat response elements. *EMBO J.* **21**, 2242–2252
17. Lee, M.S., Kliewer, S.A., Provencal, J., Wright, P.E. & Evans, R.M. (1993) Structure of the retinoid X receptor α DNA binding domain: a helix required for homodimeric DNA binding. *Science* **260**, 1117–1121
18. Mader S., Chambon, P. & White J. (1933) Defining a minimal estrogen receptor DNA binding domain. *Nucleic Acids Res.* **21**, 1125–1132
19. Claessens, F., Alen, P., Devos, A., Peeters, B., Verhoeven, G. & Rombauts, W. (1996) The androgen-specific probasin response element 2 interacts differentially with androgen and glucocorticoid receptors. *J. Biol. Chem.* **271**, 19013–19016
20. Claessens, F., Verrijdt, G., Schoenmakers, E., Haelens, A., Peeters, B., Verhoeven, G. & Rombauts, W. (2001) Selective DNA binding by the androgen receptor as a mechanism for hormone-specific gene regulation. *J. Steroid Biochem. Mol. Biol.* **76**, 23–30
21. Schoenmakers, E., Verrijdt, G., Peeters, B., Verhoeven, G., Rombauts, W. & Claessens, F. (2000) Differences in DNA binding characteristics of the androgen and glucocorticoid receptor can determine hormone specific responses. *J. Biol. Chem.* **275**, 12290–12297
22. Verrijdt, G., Schoenmakers, E., Haelens, A., Peeters, B., Verhoeven, G., Rombauts, W. & Claessens, F. (2000) Change of specificity mutations in androgen-selective enhancers. Evidence for a role of differential DNA binding by the androgen receptor. *J. Biol. Chem.* **275**, 12298–12305
23. Lefstin J.A. & Yamamoto K.R. (1998) Allosteric effects of DNA on transcriptional regulators. *Nature (London)* **392**, 885–888
24. Callewaert, L., Verrijdt, G., Christiaens, V., Haelens, A. & Claessens, F. (2003) Dual function of an amino-terminal amphipatic helix in androgen receptor-mediated transactivation through specific and non-specific elements *J. Biol. Chem.* **278**, 8212–8218
25. Christiaens, V., Bevan, C., Callewaert, L., Haelens, A., Verrijdt, G., Rombauts, W. & Claessens, F. (2002) Characterization of the two interacting surfaces of the androgen receptor and their relative role in transcriptional control. *J. Biol. Chem.* **277**, 49230–237

26. Rogatsky, I., Zarember, K.A. & Yamamoto, K.R. (2001) Factor recruitment and TIF2/GRIP1 corepressor activity at a collagenase-3 response element that mediates regulation by phorbol esters and hormones. *EMBO J.* **20**, 6071–6083
27. Black, B.E., Holaska, J.M., Rastinejad, F. & Paschal B.M. (2001) DNA binding domains in diverse nuclear receptors function as nuclear export signals. *Curr. Biol.* **11**, 1749–1758
28. Janne, O.A., Moilanen, A.M., Poukka, H., Rouleau, N., Karvonen, U., Kotaja, N., Hakli, M. & Palvimo, J.J. (2000) Androgen-receptor-interacting nuclear proteins. *Biochem. Soc. Trans.* **28**, 401–405
29. Fu, M., Wang, C., Wang, J., Sakamaki, T., Yeung, Y.G., Chang, C., Hopp, T., Fuqua, S.A., Jaffray, E., Hay, R.T., Palvimo, J.J. et al. (2002) Androgen receptor acetylation governs *trans* activation and MEKK1-induced apoptosis without affecting *in vitro* sumoylation and *trans*-repression function. *Mol. Cell. Biol.* **22**, 3373–3388
30. Thanos, D. & Maniatis, T. (1995) Virus induction of human IFN beta gene expression requires the assembly of an enhanceosome. *Cell* **83**, 1091–1100
31. Shaffer, P.L., Jivan, A., Dollins, D.E., Claessens, F. & Gewirth, D.T. (2004) Structural basis of androgen receptor binding to selective androgen response element. *Proc. Natl. Acad. Sci. U.S.A.*, in the press

6

Transcriptional activation by nuclear receptors

Mari Luz Acevedo and W. Lee Kraus[1]

Department of Molecular Biology and Genetics, Cornell University, 465 Biotechnology Building, Ithaca, NY 14853, U.S.A.

Abstract

Transcriptional activation by nuclear receptors (NRs) involves the recruitment of distinct classes of co-activators and other transcription-related factors to target promoters in the chromatin environment of the nucleus. Chromatin has a general repressive effect on transcription, but also provides opportunities for NRs to regulate transcription by directing specific patterns of chromatin remodelling and histone modification. Ultimately, the transcription of hormone-regulated genes by NRs is critically dependent on co-ordinated physical and functional interactions among the receptors, chromatin, co-activators with chromatin-, histone- and factor-modifying activities, and the RNA polymerase II transcriptional machinery. In addition, several mechanisms exist to terminate or attenuate NR-dependent signalling, including modification, recycling, subcellular redistribution and degradation of the receptors or their associated cofactors. The complexity of NR-dependent transcription provides multiple targets for regulatory inputs, thus allowing each hormone-responsive cell to direct its transcriptional output in a physiologically appropriate manner.

Introduction

Transcriptional regulation by nuclear receptors (NRs) is a complex process in which distinct classes of co-activators and other transcription-related factors

[1]*To whom correspondence should be addressed (e-mail wlk5@cornell.edu).*

are recruited to target promoters by DNA-bound receptors in the chromatin environment of the nucleus. Each factor in the ensemble contributes one or more distinct activities, such as chromatin remodelling, histone modification, cofactor complex assembly and recruitment of the basal transcription machinery, with the ultimate goal of stimulating the transcription of target genes by RNA polymerase II (RNA pol II). In this chapter, we will discuss current models for how NRs and their associated cofactors regulate the expression of hormone-regulated genes, including brief descriptions of the different types of factors and their associated activities.

Chromatin: the physiological template for NR-dependent transcription

Chromatin is the physiological template for nuclear processes involving genomic DNA, including transcriptional regulation by NRs [1]. It consists of a repeating array of DNA–protein structures called nucleosomes (Figure 1A). Chromatin is a dynamic polymer whose biochemical and biophysical properties play important and specific roles in the transcription process [2]. One effect of assembling genomic DNA into chromatin is a general repression of basal transcription [2]. Moreover, the precise positioning and subsequent mobilization of nucleosomes can regulate the access of various DNA-binding factors to the chromatin template. Furthermore, specific covalent modifications (e.g. acetylation, methylation, phosphorylation) of lysine, arginine and serine residues in the N-terminal tails of the core histones (Figure 1B) can change the properties of the nucleosome, as well as create factor-binding sites on the N-terminal tails of the histone proteins [3,4]. Distinct classes of factors recruited by NRs modify chromatin to relieve basal repression and facilitate the activation of transcription [1,5].

The RNA pol II transcriptional machinery

NRs and transcription pre-initiation complex (PIC) formation

The initiation of mRNA synthesis by RNA pol II involves the direct or indirect binding of core promoter DNA elements by a collection of 'basal' transcription factors (TFs; Figure 2A) [6,7]. The binding of ligand-activated NRs to DNA-response elements ('enhancers') in the promoter or regulatory regions of a hormone-responsive gene stimulates the assembly of a stable basal factor/RNA pol II transcription PIC at the promoter, with recognition of the TATA box and other core promoter elements by a complex called TFIID being a critical initial step [6,8]. The role of liganded NRs in promoting the formation of a stable PIC is 2-fold: (i) promoting PIC assembly through direct contacts with components of the basal transcription machinery (including TFIIB and TFIID; Figure 2B) and (ii) recruiting co-activators, which in turn promote PIC assembly through direct contacts with components of the basal transcription

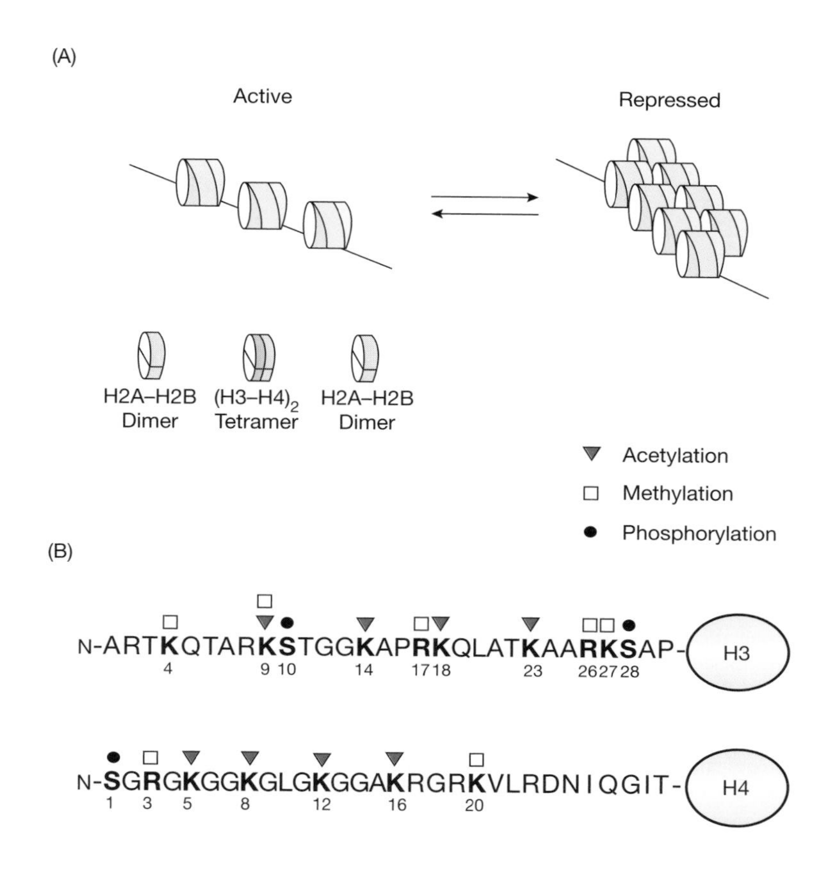

Figure 1. Chromatin structure and histone modifications
(**A**) Nucleosomes are the fundamental repeating unit of chromatin. Each nucleosome comprises two copies each of the core histones (H2A, H2B, H3 and H4), forming a histone octamer around which 1.7 turns of genomic DNA are wrapped (left-hand side). Chromatin is a dynamic polymer that can interconvert between relaxed, open and transcriptionally active conformations (left-hand side) and compact, closed and transcriptionally repressed conformations (right-hand side). (**B**) The N-terminal tails of histones H3 and H4 are covalently modified at lysine, arginine and serine residues. The specific residues and modifications are shown.

machinery and by loosening chromatin structure at the promoter [1,5,6,8]. By promoting the assembly of a stable PIC, NRs can activate transcription above the basal levels achieved with the RNA pol II machinery alone.

TATA-binding protein (TBP) and TBP-associated factors (TAFs)

TFIID is a complex of proteins containing TBP and a collection of 10–12 polypeptides called TAFs [9]. The TAFs in the TFIID complex are required for transcriptional activation by a number of different DNA-binding activators, including NRs. Several TAFs in TFIID, as well as TBP itself, make direct contacts with NRs as part of the transcriptional regulatory process [10]. For example, human $TAF_{II}30$ in TFIID binds to oestrogen receptor α (ERα),

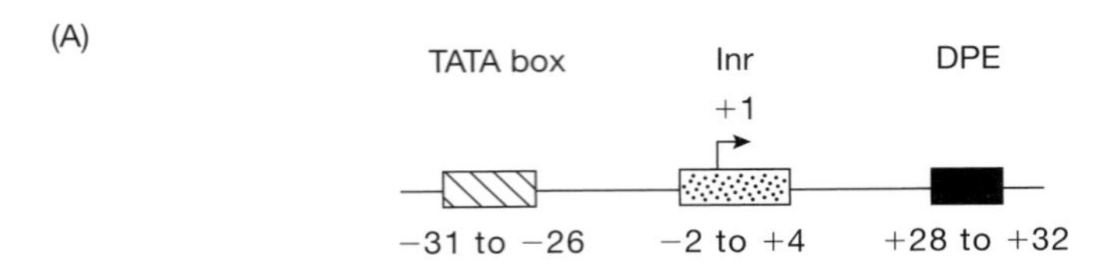

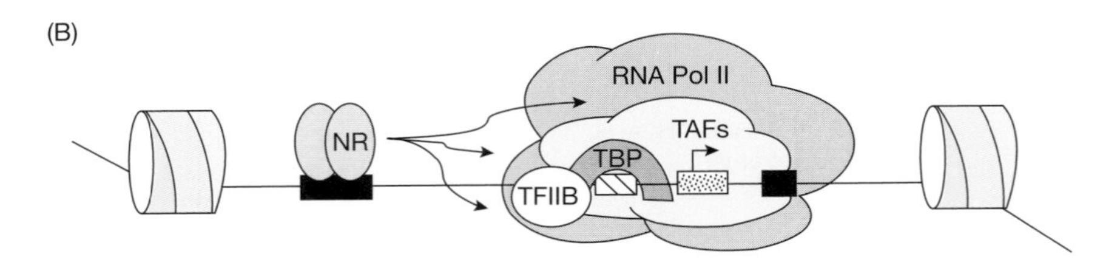

Figure 2. NRs stabilize the assembly of transcription PICs by contacting the basal transcription machinery
(**A**) Each eukaryotic promoter contains some, but not necessarily all, of the core elements shown schematically: TATA box, initiator element (Inr) and downstream promoter element (DPE). The nucleotide positions of the elements are given relative to the transcription start site (+1). (**B**) The binding of the TFIID complex [TATA-box-binding protein (TBP)+TBP-associated factors (TAFs)] to many promoters occurs via the interaction of TBP with the TATA element, an interaction that is stabilized by other basal TFs (e.g. TFIIB) and some TAFs. NRs contact a number of components of the basal transcription machinery, including TFIIB, TBP and TAFs, helping to promote the formation of a stable RNA pol II transcription PICs.

an interaction critical for ERα-dependent transcription [11]. Such interactions can help to recruit or stabilize the binding of TFIID at the promoter, a process that is enhanced by the binding of some TAFs to the core promoter elements (Figure 2B) [9]. The role of TAFs in NR-dependent transcription is illustrated by the fact that TFIID, but not TBP alone, can act synergistically with other cofactor complexes, such as Mediator and SWI/SNF, to potentiate transcription by NRs [9,12,13]. Together, the available data indicate that TAFs are required for full transcriptional activation by NRs.

NR co-activators: distinct classes with multiple activities

Co-activator properties and functions

NR co-activators comprise a large group of proteins and multipolypeptide complexes that function to enhance NR-mediated transcription. Although structurally and functionally diverse, the distinct classes of co-activators share common features [5,10,14], some of which are listed below.

- They interact with NRs directly or indirectly, typically in the presence of agonistic ligands.

- They may interact directly or indirectly with components of the basal transcription machinery (e.g. TBP, TFIIB, RNA pol II).
- They generally do not possess DNA-binding activity and have little or no transcriptional activity in the absence of a DNA-bound NR.
- They may possess enzymic activities, including kinase, acetyl-transferase, methyltransferase, ATPase and ubiquitin ligase activities.
- Multiple co-activators with distinct activities may be found in high-molecular-mass complexes.
- They may be shared by other classes of DNA-binding transcriptional activators.

In Table 1 and in the space below, we describe general classes of NR co-activators whose primary function is to promote the assembly of PICs at hormone-regulated promoters in chromatin. Other cofactors that affect NR dependent transcription through alternative mechanisms are discussed in subsequent sections.

Bridging co-activators

The regulated assembly of large multifactor NR–co-activator complexes is a critical step for NR-mediated transcription. DNA-bound NRs serve as nucleation sites for the assembly of these complexes, and a variety of additional co-activators act as bridges between NRs and their associated cofactors. Members of the steroid receptor co-activator (SRC) family of proteins (e.g. SRC-1, GRIP-1/TIF-2, ACTR/RAC3 etc., referred to as SRC-1, SRC-2 and SRC-3 under a unified nomenclature [15]) function as bridging factors between NRs and other proteins, including p300 and the highly related cAMP-response-element-binding protein (CREB)-binding protein (CBP), as well as members of the protein arginine methyltransferase (PRMT) family (including the co-activator-associated arginine methyltransferase CARM1) [10,14,15]. Like many co-activators that interact directly with NRs, SRCs possess short α-helical segments ('NR boxes') containing a sequence resembling Leu-Xaa-Xaa-Leu-Leu ('LXXLL'). NR boxes interact directly with a hydrophobic groove on the surface of NR ligand-binding domains in the presence of agonistic ligands [10,14,15]. In addition, SRCs also bind to the N-terminal region of some steroid hormone receptors, functionally bridging the N-terminal activation function 1 (AF1) with the C-terminal, ligand-regulated AF2. Separate domains of SRCs bind the non-receptor partners mentioned above [10,14,15]. Some factors, such as steroid receptor RNA activator (SRA), a functional RNA transcript with co-activator activity, promote the assembly of co-activator complexes, presumably by acting as a protein-binding scaffold [10,16]. Other factors, such as Med220 (the receptor-binding component of the Mediator co-activator complex) and Brg1/Brm (the ATPase subunits of SWI/SNF complexes), serve as a bridges between the liganded receptor and the co-activator complexes in which these subunits are contained [13,17].

Table 1. Classes of NR co-activator

CRCs, chromatin-remodelling complexes; CBP, cAMP-response-element-binding protein-binding protein; pCAF, p300/CBP-associated factor; TRAP, thyroid hormone receptor-associated protein; DRIP, vitamin D receptor-interacting protein; CARM1, co-activator-associated arginine methyltransferase 1; PRMT1, protein arginine methyltransferase 1; SRC, steroid receptor co-activator; SRA, steroid receptor RNA activator; UBE3A/E6-AP, ubiquitin protein ligase 3A/E6-associated protein; PIAS, protein inhibitor of activated STATs; STAT, signal transducer and activator of transcription; SUMO, small ubiquitin-like modifier.

Class	Examples	Functions
Basal transcription machinery-associated factors		
TAFs	Human $TAF_{II}30$	Bridge interactions between NRs and the basal transcription machinery
Mediator complexes	TRAP and DRIP	to promote the formation of stable transcription PICs at the promoter.
Bridging coactivators	SRCs and SRA	Act as scaffold to promote the assembly of co-activator complexes. Bridge interactions between DNA-bound NRs and other co-activators.
Histone- and factor-modifying enzymes		
Acetyltransferases	p300, CBP, pCAF	Covalent modification of nucleosomal histones and
Methyltransferases	CARM1, PRMT1	transcription-related factors to regulate their activities.
Kinases	Rsk2, cdk2	
Ubiquitin and SUMO ligases	UBE3A/E6-AP, PIAS	
CRCs	SWI/SNF and related Brg1- and Brm-containing complexes	ATP-dependent disruption of nucleosomal structures to increase DNA accessibility.

Co-activators with histone- and factor-modifying activities

Post-translational modifications can have dramatic effects on the biochemical activities of proteins; stimulatory in some cases, inhibitory in others. The covalent modification of core histone proteins is now recognized as a critical epigenetic regulator of transcription and other chromatin-dependent processes [3,4]. A number of well-characterized co-activators possess intrinsic enzymic activities capable of covalently modifying the N-terminal tails of core histones and linker histones [5,10,14]. The enzymic activities include histone acetyltransferases [HATs; e.g. p300, CBP and pCAF (p300/CBP-associated factor)], histone methyltransferases (e.g. PRMT1 and CARM1) and histone kinases (such as cdk2, which can phosphorylate histone H1). The co-ordinated actions of these histone-modifying enzymic activities, which in many cases are regulated as endpoints in cellular signalling pathways, specify a 'histone code' at the promoter [3,4]. In the case of acetylation and phosphorylation, partner enzymes (histone deacetylases and phosphatases) can remove the modification, thus reversing the effect. In addition to their actions on histones, a number of co-activators with enzymic activity (e.g. p300, pCAF, CARM1) have also been shown to covalently modify non-histone chromatin-associated proteins (e.g. high mobility group proteins) and transcription-related factors (e.g. co-activators such as SRC-3 and NRs such as androgen receptor), thereby altering the activities of those factors [5,18]. Although the enzymic activities of the histone-modifying co-activators are critical for NR-dependent transcription, they are typically not sufficient for co-activator function. Other domains in these proteins may contribute 'classical' co-activator functions, such as binding to RNA pol II [5,10].

Chromatin-remodelling complexes (CRCs)

The packaging of genomic DNA into nucleosomes restricts NR-dependent assembly of transcription complexes at target promoters, in spite of the fact that many NRs bind stably and with high affinity to their cognate DNA-response elements in chromatin. NRs promote the formation of an open chromatin architecture at promoters through the ligand-dependent recruitment of CRCs that interact directly with the receptors via one or more subunits [5,8,17]. CRCs are multipolypeptide enzymes categorized by the type of ATPase subunit that they contain (e.g. human SWI/SNF represents a family of related complexes containing either Brg1 or Brm as the ATPase subunit) [19]. The exact composition of the complexes can vary from one cell type to the next, allowing for greater functional flexibility [12,17,19]. CRCs use the energy stored in ATP to mobilize or structurally alter nucleosomes, allowing greater access of the transcriptional machinery to promoter DNA, thus facilitating transcriptional activation [19]. Although ATP-dependent chromatin remodelling is required for NR-dependent transcription, it is not sufficient. The actions of CRCs may set the stage for subsequent actions by co-activators with histone-modifying activities [8,19,20].

Mediator complexes

Interactions between DNA-bound NRs and the RNA pol II transcriptional machinery help to promote the formation of stable transcription PICs at the promoter. The multipolypeptide Mediator complexes ('Mediator'; also known as the TRAP and DRIP complexes) are a class of NR- and RNA pol II-interacting co-activators that serve such a function [13,21]. At least two individual subunits of Mediator can interact directly with NRs, as demonstrated by a variety of biochemical studies [13,21]. Med220 binds to NR ligand-binding domains in a ligand-dependent manner via a receptor interaction domain that contains two NR boxes. This subunit is responsible for the association of the entire Mediator complex with a variety of NRs *in vitro* and is probably responsible for the recruitment of Mediator to the promoters of NR-regulated genes after ligand induction *in vivo*. In contrast, Med150 binds to the N-terminal region of some steroid hormone receptors without a requirement for ligand. As discussed below, Mediator has multiple specific functions, including promoting the re-assembly of PICs during the transcription process [13,21].

Histone modifications and the histone code

The role of histone modifications in the transcription of genes in chromatin

A link between covalent post-translational histone modifications and transcriptional regulation has been well established [1,3,4]. Current results suggest that histone modifications play multiple distinct roles in transcriptional regulation with chromatin templates and, as such, can dramatically affect NR-dependent transcription. For example, histone acetylation (i) facilitates chromatin remodelling by 'loosening' histone–DNA contacts within nucleosomes and histone–histone contacts between nucleosomes through charge neutralization, and (ii) alters the structure of nucleosomes by promoting the transfer of histone H2A–H2B dimers from nucleosomes to histone chaperones [1,3,5]. Furthermore, specific patterns of acetylation and methylation can also create factor-binding sites on the N-terminal tails of core histones [3,4]. Factors containing histone-binding domains (e.g. bromodomains and chromodomains), including the NR co-activators p300 and CBP which contain bromodomains, specifically recognize and bind to the modified histones. Thus the pattern of covalent histone modification specifies a histone code that dictates which factors associate directly with chromatin at a particular promoter [3,4]. The recruitment of cofactors by DNA-bound NRs may dictate gene specificity for the transcriptional response, whereas direct interactions of cofactors with chromatin may provide additional mechanisms for stabilizing complex formation at the promoter.

Specificity of histone-modifying enzymes and regulatory crosstalk

A question relevant to our thinking about histone modification is why so many different histone-modifying enzymes are required by NRs to activate the transcription of genes in chromatin. The answer lies in the substrate specificities of each enzyme and the histone code described above. Each histone-modifying enzyme can modify a distinct set of targets, ultimately directing a distinct set of outcomes. For example, the HATs p300 and pCAF have different specificities for histone proteins; p300 preferentially acetylates H3-K14/18 and H4-K5/8, whereas pCAF primarily acetylates H3-K14 (Figure 1B) [5]. The partially overlapping, rather than identical, substrate specificities for individual histone-modifying enzymes within each class of enzyme (e.g. HATs) may underlie the transcriptional synergism observed between members of each class during the stimulation of NR activity [3,5].

Functional interplay between HATs and histone methyltransferases has also been observed at both the enzymic and transcriptional levels [3–5]. For example, methylation of H4-R3 by PRMT1 increases acetylation of H4-K8 and K12 by p300. In contrast, pre-acetylation of H4 reduces subsequent H4-R3 methylation by PRMT1 (Figure 1B). Interestingly, phosphorylation of H3-S10 is tied to intracellular protein kinase signalling pathways and can enhance the subsequent acetylation of nearby lysine residues. Although H3 phosphorylation has not yet been shown to play a role in NR-dependent transcription with chromatin, it is interesting to note that some of the same signalling pathways that enhance H3-S10 phosphorylation (e.g. mitogen-activated protein kinase pathways) have been shown to enhance NR transcriptional activity [5]. These results illustrate some of the complex interactions that can occur between different factors with histone-modifying activities, generating interplay at the level of the nucleosome that results in a 'combinatorial histone code' [3,4].

Cofactor recruitment and activity at NR-regulated promoters

Cofactor recruitment and dynamics

The precise timing of the recruitment of cofactors and their associated activities to promoters by DNA-bound NRs is critical for the appropriate regulation of gene expression [5,19,20]. Much of the current research has focused on understanding the ordered recruitment of HAT complexes and CRCs, leading to the co-ordination of histone acetylation and nucleosome remodelling. The combined data from chromatin immunoprecipitation ('ChIP') assays and *in vitro* 'order-of-factor-addition' transcription experiments have uncovered two pathways for the recruitment of HATs and CRCs to eukaryotic promoters (Figure 3) [19,20]. In the first, HATs are recruited initially and facilitate the subsequent recruitment of CRCs, possibly by creating binding sites on histone tails for bromodomain-containing CRCs.

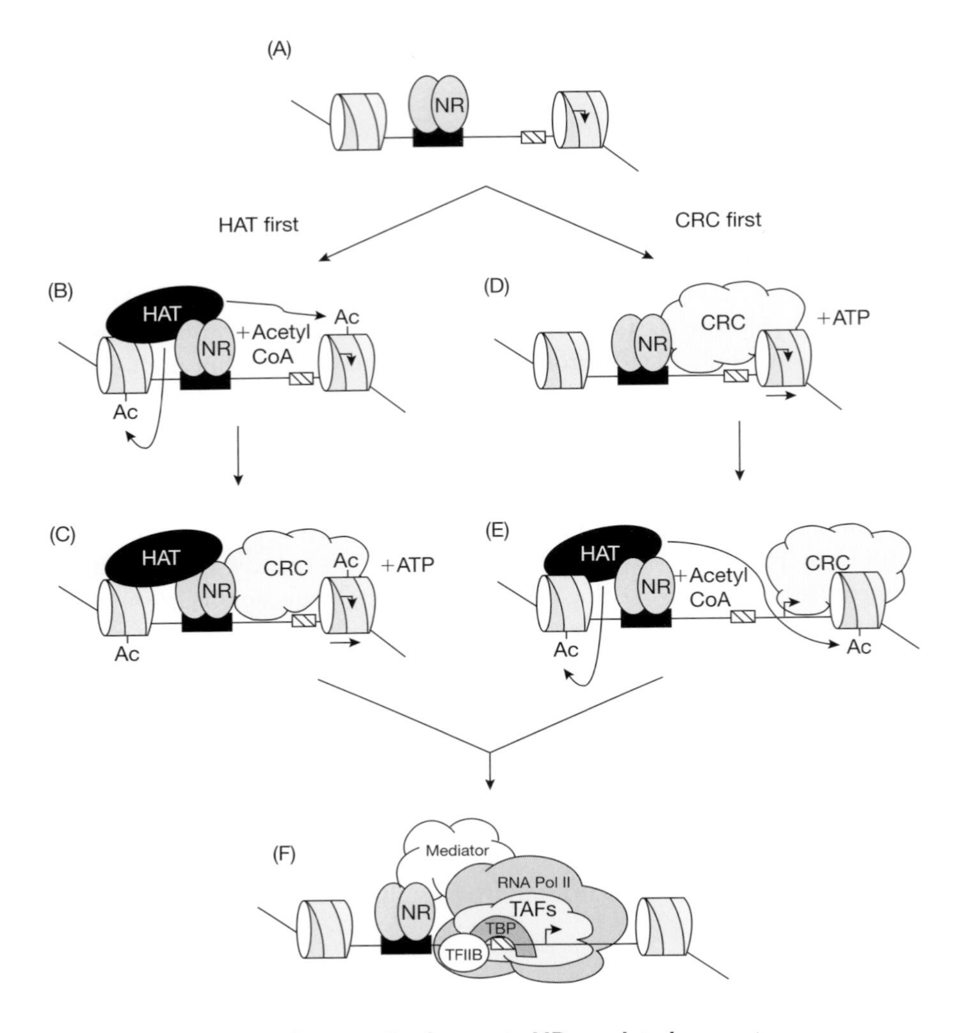

Figure 3. Ordered recruitment of cofactors to NR-regulated promoters
The available data suggest two pathways for the recruitment of HATs and CRCs to eukaryotic promoters [19,20]. In the first, HATs are recruited initially and facilitate the subsequent recruitment of CRCs (pathway A, B, C, F). In the second, CRCs are recruited initially and facilitate the subsequent recruitment of HATs (pathway A, D, E, F). Mediator and components of the basal machinery enter at various points in the transcription process, in many cases subsequent to histone modification and chromatin remodelling (as depicted here). Variations in the order of recruitment of the factors shown are possible and are likely to be dependent on specific promoter sequences and chromatin architectures.

In the second, CRCs are recruited initially and facilitate the subsequent recruitment of HAT complexes, presumably by promoting a more accessible chromatin architecture that allows for NR-dependent recruitment of the HATs. Mediator and components of the basal machinery enter at various points in the transcription process, in many cases subsequent to histone modification and chromatin remodelling [8,19,20]. Overall, the order in which

HATs, CRCs, and other factors are recruited is determined by the structure of the promoter (i.e. the DNA elements it contains, its chromatin architecture) and has important consequences for determining the mechanisms of gene-specific transcriptional regulation [19,20].

Although a number of ChIP studies have examined the recruitment of subsets of factors to NR-regulated promoters (see, for example [22–24]), the limited number of genes, factors and NRs simultaneously examined in each study makes it difficult to establish a comprehensive or general model for PIC assembly at NR-regulated promoters *in vivo*. However, a few key points are worth noting. First, the initial association of NRs, cofactors and RNA pol II with hormone-regulated genes occurs rapidly following hormone treatment (within 15–30 min) [22,23]. Second, in some cases, cofactors may cycle on and off the promoter during the transcription process [22,24]. Third, cofactors with overlapping but distinct activities (e.g. p300, CBP and pCAF) may be recruited at different points during the transcription process [5,22,23]. Fourth, cofactors that compete for binding to liganded NRs (e.g. SRCs and Mediator) may associate concurrently with promoters [22], raising an unresolved issue of how multiple co-activators with similar receptor-binding specificities are recruited simultaneously by the ligand-bound receptor. Finally, cofactor and RNA pol II dissociation occurs simultaneously with the attenuation of the transcriptional response [23,25].

Co-ordination of co-activator-specific activities during the transcription cycle

Each cofactor recruited by a NR to a previously unstimulated promoter provides a particular activity (or activities) required for a specific step in the transcription process. For example, co-activators with histone- and chromatin-modifying activities are required for the formation of an open chromatin architecture at the promoter conducive to the formation of a stable PIC and the initial burst of transcription [26,27]. These same co-activators may not be needed for subsequent rounds of transcription reinitiation, promoter clearance and elongation from the same promoter, presumably because the chromatin architecture remains in an open conformation. In contrast, Mediator, which helps to promote the stable association of RNA pol II with the promoter, may be required in each round of transcription to promote the formation of a stable PIC [13,27]. Recent studies suggest that NRs and their associated cofactors can also influence various post-initiation RNA processing events, including intron skipping and retention, 5′ splice site selection and polyadenylation [28]. These examples highlight only a few of the many biochemical activities provided by co-activators at specific times during the transcription cycle. Importantly, due to the complementary nature of the distinct activities provided by different classes of co-activators during the transcription process, transcriptional synergism among co-activators is frequently observed [5,8,21].

Shutting off NR-dependent transcription: factor modification, recycling, redistribution and turnover

The transcription of inducible genes is tightly regulated in order to maintain proper patterns of gene expression in response to a variety of physiological stimuli. Thus several mechanisms exist to terminate the transcription of NR-regulated genes (Figure 4). For example, recent studies have shown that chaperone proteins, such as p23 and possibly heat-shock protein 90 (hsp90), can promote the removal of NRs and RNA pol II from DNA, perhaps as part of a dynamic cycle of transcription complex disassembly and re-formation that ultimately leads to the attenuation of transcription [25]. Such a mechanism may allow for continuous sampling of hormone levels by NRs, allowing reassembly of the transcription complexes only when hormone levels remain appropriately elevated.

Another mechanism for shutting off transcription involves the covalent post-translational modification of co-activators (e.g. lysine acetylation, arginine methylation), which can inhibit the binding of co-activators to NRs or other transcriptional activators by chemically altering critical protein–protein interaction surfaces [5,18]. For example, the acetylation of SRC-3 by p300 has been shown to cause a disruption of receptor–co-activator complexes, leading to a decrease in receptor-mediated gene activation [23]. Some NRs, including

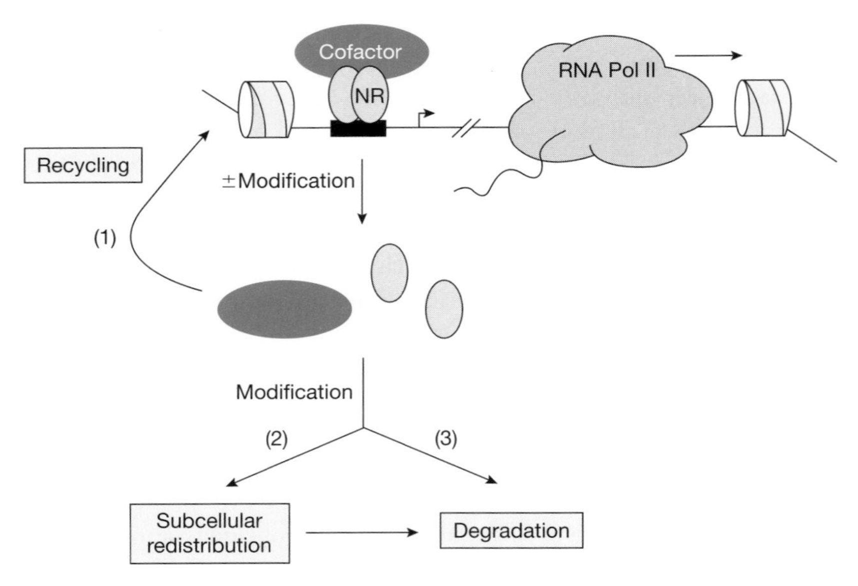

Figure 4. Reduce, redistribute, recycle: shutting off NR-dependent transcription
Several mechanisms exist to attenuate NR-regulated transcription. After dissociation from the promoter (a process that may, in some cases, be promoted by covalent modification), NRs and their associated cofactors can: (1) recycle back to the promoter to stimulate subsequent rounds of transcription, (2) be subjected to covalent modifications that target them for subcellular redistribution or (3) be subjected to covalent modifications that target them for degradation.

the androgen receptor, are also acetylated by co-activators with acetyltransferase activity, such as p300 [29]. The consequences of these NR modifications have not yet been fully explored, but they appear to modify receptor activity.

After dissociation of transcription complexes, the constituent factors may be redistributed to different subcellular compartments. Some co-activators, such as SRC-3 and Trip230, exhibit nuclear localization (and co-localization with NRs) only in the presence of an appropriate cellular signal which may, in some cases, result in the covalent post-translational modification of the co-activator (e.g. phosphorylation) [30,31]. Thus redistribution to non-nuclear (or in some cases subnuclear) compartments can sequester co-activators away from NRs and the transcriptional machinery, or perhaps decrease their activities, thus leading to an attenuation of the transcriptional response after the signal is removed.

Factor turnover involving the ubiquitin-dependent proteasome pathway represents an additional mechanism for the attenuation of NR-dependent transcriptional responses (Figure 4) [24,32,33]. The covalent attachment of ubiquitin, a 76-amino-acid peptide, to NRs and co-activators by ubiquitin-conjugating enzymes marks these factors for recognition and degradation by the 26 S proteasome, a multisubunit protease [32,33]. In many cases, down-regulation of receptors and co-activators results in a parallel reduction in the transcription of NR-target genes, preventing overstimulation by hormonal signals. However, in other cases, proteasome actions are required for maximum transcription of NR-target genes, suggesting that NR and co-activator turnover may also act to reset the transcriptional apparatus in preparation for a subsequent transcriptional response [32,33]. Interestingly, but perhaps not surprisingly, some components of the ubiquitin [and SUMO (small ubiquitin-like modifier), a related modification] conjugating systems have been identified as NR-interacting proteins [10,32] (Table 1).

Cell-type-specificity and co-integration of multiple signalling pathways

Limiting hormonal responses to specific tissues or cell types is an important aspect of signalling, especially when the hormone is released systemically. With regard to NRs and their associated cofactors, cell-type-specific actions are best illustrated by the wide variety of phenotypes evident in gene-knockout studies in mice [34]. Cell-type specificity is generally established in two ways: (i) by limiting the expression of a particular NR or cofactor to a certain cell type and (ii) by involving multiple distinct NR or cofactor subtypes in the response to a single hormone. These two possibilities are not mutually exclusive, as different subtypes of a particular NR or cofactor may exhibit cell-type-specific expression in target tissues. The complement of NRs and cofactors in a cell can have profound effects on signalling outcomes and may explain how the same receptor proteins can elicit distinct responses to a ligand in different cell types. For example, in endometrial cells, which have high levels of SRC-1, the synthetic ER ligand tamoxifen acts as an agonist of ER-dependent endpoints, but in mammary cells, which have low levels of SRC-1, tamoxifen acts as an antagonist [35]. With some multisubunit co-activator complexes (e.g. SWI/SNF and related CRCs), the cell-type-specific expression of certain subunits can change the activity of complex with NRs [17]. Finally, one must consider the fact that a single cell may possess multiple NR and non-NR signalling pathways that operate simultaneously. Many of these pathways converge on a shared set of cofactors, allowing for co-integration of signal-regulated transcription [14,34]. In some cases, competition for limiting shared cofactors may allow the co-integration of transcriptional outcomes [14]. Together, cell-type specificity and co-integration may increase the complexity of NR signalling, but they also increase the opportunities for regulation and control.

Conclusion

Transcriptional regulation by NRs is a complex process involving a wide array of cofactors, each possessing a distinct activity or activities. The complexity of the transcriptional process provides many opportunities for regulation beyond the initial hormone-dependent, NR-regulated event. Such regulatory plasticity ultimately allows each cell to direct its transcriptional output in a physiologically appropriate manner. Although much progress has been made in understanding this process over the past decade, many questions still remain. The future should bring a greater understanding of the molecular mechanisms by which multiple signalling pathways converge on a single NR, the extent and importance of tissue-specific responses, and insight into the role of higher order chromatin structures in gene regulation.

Summary

- *Chromatin is the physiological template for NR-mediated transcription.*
- *The assembly of promoters into chromatin has a general repressive effect on transcription, but also provides opportunities for regulating transcription through chromatin remodelling and the histone-modification code.*
- *NRs stimulate the transcription of hormone-regulated genes by recruiting the basal transcription machinery and promoting the formation of stable RNA pol II PICs.*
- *NRs serve as nucleation sites for the recruitment of distinct classes of co-activator to promoters in chromatin.*

- *Some co-activators possess histone- and chromatin-modifying activities that alter nucleosome positioning and chromatin structure, as well as create factor-binding sites on the N-terminal tails of the core histones. Other co-activators interact with both NRs and the transcriptional machinery to promote the stable binding of RNA pol II at the promoter.*
- *The precise timing of the recruitment of cofactors to promoters in chromatin is determined by the DNA and chromatin structure of the promoter and has important consequences for determining the mechanisms of gene-specific transcriptional regulation.*
- *Several mechanisms exist to terminate the transcription of NR-regulated genes. They generally involve modification, recycling, subcellular redistribution, and degradation of the receptors or their associated cofactors.*
- *Cell-type-specific activity of NRs, which can have profound effects on signalling outcomes, is established by limiting the expression of a particular NR or cofactor (or subtypes) to certain cells.*

We thank members of the Kraus Lab for critical reading of this review. We also thank the Burroughs Wellcome Fund, the U.S. National Institutes of Health NIDDK, the American Cancer Society and the Komen Breast Cancer Foundation for support of research that led to some of the results and ideas presented.

References

1. Urnov, F.D. & Wolffe, A.P. (2001) A necessary good: nuclear hormone receptors and their chromatin templates. *Mol. Endocrinol.* **15**, 1–16
2. Horn, P.J. & Peterson, C.L. (2002) Chromatin higher order folding – wrapping up transcription. *Science* **297**, 1824–1827
3. Jenuwein, T. & Allis, C.D. (2001) Translating the histone code. *Science* **293**, 1074–1080
4. Fischle, W., Wang, Y. & Allis, C.D. (2003) Histone and chromatin cross-talk. *Curr. Opin. Cell Biol.* **15**, 172–183
5. Kraus, W.L. & Wong, J. (2002) Nuclear receptor-dependent transcription with chromatin. Is it all about enzymes? *Eur. J. Biochem.* **269**, 2275–2283
6. Orphanides, G., Lagrange, T. & Reinberg, D. (1996) The general transcription factors of RNA polymerase II. *Genes Dev.* **10**, 2657–2683
7. Butler, J.E. & Kadonaga, J.T. (2002) The RNA polymerase II core promoter: a key component in the regulation of gene expression. *Genes Dev.* **16**, 2583–2592
8. Dilworth, F.J. & Chambon, P. (2001) Nuclear receptors coordinate the activities of chromatin remodeling complexes and coactivators to facilitate initiation of transcription. *Oncogene* **20**, 3047–3054
9. Albright, S.R. & Tjian, R. (2000) TAFs revisited: more data reveal new twists and confirm old ideas. *Gene* **242**, 1–13
10. McKenna, N.J., Lanz, R.B. & O'Malley, B.W. (1999) Nuclear receptor coregulators: cellular and molecular biology. *Endocr. Rev.* **20**, 321–344
11. Jacq, X., Brou, C., Lutz, Y., Davidson, I., Chambon, P. & Tora, L. (1994) Human TAFII30 is present in a distinct TFIID complex and is required for transcriptional activation by the estrogen receptor. *Cell* **79**, 107–117
12. Lemon, B., Inouye, C., King, D.S. & Tjian, R. (2001) Selectivity of chromatin-remodelling cofactors for ligand-activated transcription. *Nature (London)* **414**, 924–928

13. Malik, S. & Roeder, R.G. (2000) Transcriptional regulation through Mediator-like coactivators in yeast and metazoan cells. *Trends Biochem. Sci.* **25**, 277–283
14. Glass, C.K. & Rosenfeld, M.G. (2000) The coregulator exchange in transcriptional functions of nuclear receptors. *Genes Dev.* **14**, 121–141
15. Leo, C. & Chen, J.D. (2000) The SRC family of nuclear receptor coactivators. *Gene* **245**, 1–11
16. Lanz, R.B., McKenna, N.J., Onate, S.A., Albrecht, U., Wong, J., Tsai, S.Y., Tsai, M.J. & O'Malley, B.W. (1999) A steroid receptor coactivator, SRA, functions as an RNA and is present in an SRC-1 complex. *Cell* **97**, 17–27
17. Hsiao, P.W., Deroo, B.J. & Archer, T.K. (2002) Chromatin remodeling and tissue-selective responses of nuclear hormone receptors. *Biochem. Cell Biol.* **80**, 343–351
18. Freiman, R.N. & Tjian, R. (2003) Regulating the regulators: lysine modifications make their mark. *Cell* **112**, 11–17
19. Narlikar, G.J., Fan, H.Y. & Kingston, R.E. (2002) Cooperation between complexes that regulate chromatin structure and transcription. *Cell* **108**, 475–487
20. Cosma, M.P. (2002) Ordered recruitment: gene-specific mechanism of transcription activation. *Mol. Cell* **10**, 227–236
21. Rachez, C. & Freedman, L.P. (2001) Mediator complexes and transcription. *Curr. Opin. Cell Biol.* **13**, 274–280
22. Shang, Y., Hu, X., DiRenzo, J., Lazar, M.A. & Brown, M. (2000) Cofactor dynamics and sufficiency in estrogen receptor-regulated transcription. *Cell* **103**, 843–852
23. Chen, H., Lin, R.J., Xie, W., Wilpitz, D. & Evans, R.M. (1999) Regulation of hormone-induced histone hyperacetylation and gene activation via acetylation of an acetylase. *Cell* **98**, 675–686
24. Reid, G., Hübner, M.R., Métivier, R., Brand, H., Denger, S., Manu, D., Beaudouin, J., Ellenberg, J. & Gannon, F. (2003) Cyclic, proteasome-mediated turnover of unliganded and liganded ER α on responsive promoters is an integral feature of estrogen signaling. *Mol. Cell* **11**, 659–707
25. Freeman, B.C. & Yamamoto, K.R. (2002) Disassembly of transcriptional regulatory complexes by molecular chaperones. *Science* **296**, 2232–2235
26. Kraus, W.L. & Kadonaga, J.T. (1998) p300 and estrogen receptor cooperatively activate transcription via differential enhancement of initiation and reinitiation. *Genes Dev.* **12**, 331–342
27. Acevedo, M.L. & Kraus, W.L. (2003) Mediator and p300/CBP-steroid receptor coactivator complexes have distinct roles, but function synergistically, during estrogen receptor α-dependent transcription with chromatin templates. *Mol. Cell Biol.* **23**, 335–348
28. Auboeuf, D., Honig, A., Berget, S.M. & O'Malley, B.W. (2002) Coordinate regulation of transcription and splicing by steroid receptor coregulators. *Science* **298**, 416–419
29. Fu, M., Wang, C., Wang, J., Zhang, X., Sakamaki, T., Yeung, Y.G., Chang, C., Hopp, T., Fuqua, S.A., Jaffray, E. et al. (2002) Androgen receptor acetylation governs trans activation and MEKK1-induced apoptosis without affecting *in vitro* sumoylation and trans-repression function. *Mol. Cell Biol.* **22**, 3373–3388
30. Chen, Y., Chen, P.L., Chen, C.F., Sharp, Z.D. & Lee, W.H. (1999) Thyroid hormone, T3-dependent phosphorylation and translocation of Trip230 from the Golgi complex to the nucleus. *Proc. Natl. Acad. Sci. U.S.A.* **96**, 4443–4448
31. Qutob, M.S., Bhattacharjee, R.N., Pollari, E., Yee, S.P. & Torchia, J. (2002) Microtubule-dependent subcellular redistribution of the transcriptional coactivator p/CIP. *Mol. Cell Biol.* **22**, 6611–6626
32. Dennis, A.P., Haq, R.U. & Nawaz, Z. (2001) Importance of the regulation of nuclear receptor degradation. *Front. Biosci.* **6**, D954–D959
33. Lonard, D.M., Nawaz, Z., Smith, C.L. & O'Malley, B.W. (2000) The 26S proteasome is required for estrogen receptor-α and coactivator turnover and for efficient estrogen receptor-α transactivation. *Mol. Cell* **5**, 939–948
34. Curtis, S.H. & Korach, K.S. (2000) Steroid receptor knockout models: phenotypes and responses illustrate interactions between receptor signaling pathways *in vivo*. *Adv. Pharmacol.* **47**, 357–380
35. Shang, Y. & Brown, M. (2002) Molecular determinants for the tissue specificity of SERMs. *Science* **295**, 2465–2468

7

Gene repression by nuclear hormone receptors

Udo Moehren, Maren Eckey and Aria Baniahmad[1]

Genetic Institute, Justus-Liebig-University, 35392 Giessen, Germany

Abstract

Repression by nuclear hormone receptors (NHRs) plays an important role in development, immune response and cellular function. We review mechanisms of how NHRs act as repressors of gene transcription either by direct contact with basal transcription factors or through recruitment of cofactors and enzymic activities that modulate chromatin accessibility. We describe also the role and biochemical mechanism of the cognate hormone that switches a NHR from a transcriptional silencer into an activator. This includes data from crystal structure, functional receptor domain analyses and the role of co-repressors in chromatin modification and remodelling. Furthermore, the comparison of negative response elements with classical response elements unravels the role of co-repressors in this context. We also describe the inhibition of the nuclear factor κB and Jun/Fos pathway by NHRs, as well as the molecular mechanism of anti-hormone therapies. Anti-hormones are commonly used in breast and prostate cancer therapy to inhibit cancer proliferation through repression of the oestrogen or androgen receptor, respectively. Here we provide a comprehensive overview of the various mechanism of NHR repression.

1To whom correspondence should be addressed (e-mail Aria.Baniahmad@gen.bio.uni-giessen.de).

Introduction

More and more studies are revealing that gene silencing is just as important as gene activation. Through these studies many interesting new aspects of the molecular action of gene control have emerged. Nuclear hormone receptors (NHRs) represent a large family of ligand-regulated transcription factors. They can bind to their cognate hormone-response elements on DNA and regulate, activate or repress the expression of their target genes. They can be roughly divided into receptors that bind to steroids, receptors that bind to non-steroids and orphan receptors for which no ligand is yet known.

Interestingly, hormone-bound NHRs can have both activation and repression functions. When bound to their response element they activate target genes; however, they can also repress genes through inhibition of other signalling pathways, which is termed crosstalk [1]. Unliganded NHRs act either as gene silencers or are transcriptionally inactive. Unliganded receptors for steroids, such as the androgen receptor (AR), are localized in the cytoplasm and become activated by hormone, which leads to dissociation of heat-shock factors from the receptor and translocation to the nucleus [2]. In contrast, unliganded non-steroid NHRs such as thyroid hormone receptor (TR), retinoic acid receptor (RAR) or most orphan receptors are located in the nucleus, where they bind to their response elements. These non-steroid NHRs silence target genes in an active manner by recruitment of co-repressors [3]. Gene silencing by NHRs is a very important and crucial function *in vivo* since aberrant silencing leads to disease and developmental abnormalities. In this chapter we review the different levels and mechanisms of repression by NHRs.

Gene-silencing by co-repressors

NHRs have a tripartite structure. The DNA-binding domain is highly conserved, consists of two zinc-finger motifs, and separates the receptor N-terminus from its C-terminus. Both TRs and RARs silence gene transcription

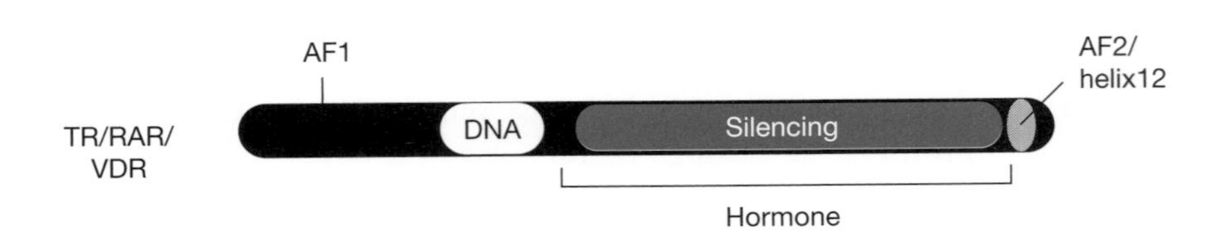

Figure 1. Functional domains of nuclear receptors with silencing domains
Schematic representation of the common functional domains of the TRs, RARs and the vitamin D receptor (VDR). The DNA-binding domain (DNA) consists of the classical nuclear receptor two-zinc-finger domain. The location and expansion of the hormone-binding and silencing domain is indicated. The C-termini of most nuclear receptors consist of 12 helices. Helix 12 is required for hormone-induced co-repressor dissociation and hormone-dependent transactivation.

in the absence of hormone through a silencing domain (Figure 1) in the receptor's C-terminus, a region overlapping the ligand-binding domain (LBD) [2]. One mechanism of repression is through binding of co-repressors. Co-repressors are non-DNA-binding cofactors, which are located in the nucleus and harbour an autonomous silencing domain [3]. These factors interact with the silencing domain in a hormone-sensitive manner. Binding of the cognate hormone induces a conformational change in the receptor resulting in dissociation of co-repressors and relief of gene silencing [2]. In the following we summarize characteristics of selected co-repressors for NHRs focusing on the best characterized co-repressor class, SMRT (silencing mediator for RAR and TR)/NCoR (nuclear receptor co-repressor), which can be considered as a paradigm for co-repressor action.

Various co-repressors for NHR have been identified that can be divided by their mode of interaction with NHR. Co-repressors such as SMRT, NCoR, Alien, Hairless (Hr) and SMRTER (SMRT-related ecdysone receptor-interacting factor) interact in a ligand-sensitive manner with NHR. Ligand binding leads to their dissociation from the receptor. SUN-CoR (small unique nuclear receptor co-repressor) and PSF (polypyrimidine tract binding protein-associated splicing factor) are co-repressors that bind constitutively to NHR, while RIP140 (receptor-interacting protein 140) and LCoR (ligand-dependent nuclear receptor co-repressor) bind only to liganded receptors [4–8]. Here, we briefly provide a few characteristics of these co-repressors.

SMRT/NCoR

The co-repressors first characterized for NHRs are SMRT and NCoR, which bind to the silencing domains of TRs, RARs and other NHRs only in the absence of ligand [9]. NCoR and SMRT show very similar domain structures and functionality, and can be separated into a functional C-terminal part interacting with NHRs and an N-terminal part which mediates the repression signal to both the basal transcription machinery and to chromatin. Two specific nuclear receptor interaction domains (IDs) within SMRT/NCoR have been identified. In analogy to the Leu-Xaa-Xaa-Leu-Leu (LXXLL) sequence motifs found in co-activators (see Chapter 6 in this volume), similar hydrophobic motifs, termed ΦXXΦΦ (where Φ is any hydrophobic amino acid) or CoRNR (co-repressor/nuclear receptor box) motifs in the IDs, have been shown to mediate interactions with NHRs [10]. It is discussed that each ID contacts one nuclear receptor LBD within the NHR dimers. Interestingly, the IDs differ in their affinities for specific receptors. RAR and TR, for example, have higher affinities for ID1, whereas RXR (retenoid X receptor) shows a preference for ID2 [11]. Notably, recent evidence shows that NCoR is more relevant for TR-mediated gene silencing as SMRT [12].

Overexpression of SMRT increases receptor-mediated silencing in the absence of hormone. Both NCoR and SMRT are complexed with other proteins that have histone deacetylase (HDAC) activity, which is thought to render chromatin more condensed and therefore becomes less accessible. In addition, TR-, SMRT- and NCoR-mediated silencing are, at least in part, sensitive to the HDAC inhibitor trichostatin A. This indicates that NHRs with silencing function recruit HDAC activity for target gene repression.

The crystal structures of LBDs from different NHRs reveal a structure that consists of 12 α-helices. A region within Helix 1, the CoR box, is crucial for the recruitment of SMRT and NCoR, while a binding surface formed by Helices 3, 4 and 5 is needed for co-repressor as well as co-activator binding [11]. A key player in the regulation of the hormone-sensitive interaction is Helix 12 (Figures 1 and 2). In the unliganded state, Helix 12 protrudes away from the core structure of the LBD. However, ligand binding induces a rotation of Helix 12 towards the LBD core, now forming a lid on the hormone-binding cavity. Amino acid substitutions that abrogate this conformational change can impair co-repressor release after hormone binding [13]. In addition, NHRs that lack Helix 12 have constitutive silencing function, despite binding ligand [14]. This strongly suggests that Helix 12 regulates co-repressor release and association of co-activators upon hormone binding.

SMRT and NCoR mediate the repression function by at least three transferable repression domains, which show no homology with each other. This fact suggests several non-redundant and simultaneous modes of repression by SMRT/NCoR which are discussed below [11].

In *Drosophila*, SMRTER, a functional analogue of SMRT/NCoR, has been identified. Failure of ecdysone receptor to silence genes and to bind SMRTER leads to developmental defects [15].

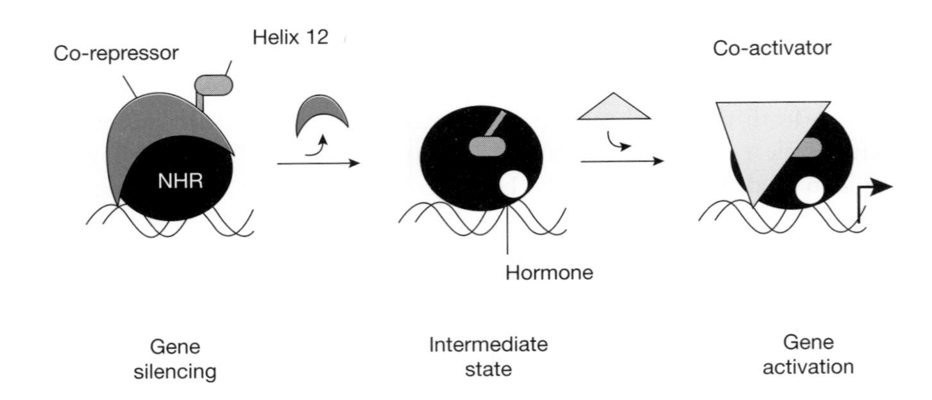

Figure 2. Molecular switch of nuclear receptor function through hormone binding
Mechanisms of gene silencing and hormone-dependent transcriptional activation are mediated through differential cofactor binding. Members of NHRs are associated with co-repressors. Upon cognate hormone binding (white circle) Helix 12 changes its position and moves towards the receptor and binds to the holo-receptor body. This leads to co-repressor dissociation, loss of silencing, co-activator recruitment and subsequent gene activation. Arrows indicate strength of promoter activity.

Alien

Alien shows a hormone-sensitive interaction with TR [16–18]. Interestingly, Alien exhibits different receptor specificity from the SMRT/NCoR class due to its lack of interaction with RAR or RXR. Alien interacts in the absence of ligand with TR, the vitamin D receptor and with the orphan receptor DAX-1. Alien shows no significant homologies to the SMRT/NCoR class but harbours ΦXXΦΦ motifs. Interestingly, Alien is highly conserved, showing 90% identity between human and *Drosophila*, indicating an essential evolutionarily conserved function. Similarly, *Drosophila* Alien interacts specifically with *Drosophila* NHRs, such as ecdysone receptor/Svp (seven-up) [16]. Recently, it was shown that the expression of Alien is controlled by thyroid hormone in the developing rat brain and in cultured cells, suggesting a negative-feedback mechanism between TR and its own co-repressor. Thus the reduction of co-repressor levels may represent a control mechanism of TR-mediated gene silencing [19].

Hr

Hr was recently also identified as a co-repressor for TR, using HDAC activity for repression. Interestingly, Hr shows tissue- and development-specific expression and, similar to Alien, its expression is also controlled by thyroid hormone in brain [20].

SUN-CoR

SUN-CoR which exhibits hormone-independent interaction with NHRs may modulate both ligand and unliganded NHR action [4].

PSF

Similarly, PSF is a co-repressor that binds hormone-independently to TR and RXR, mediating repression through recruitment of Sin3A. In contrast with SMRT/NCoR, Alien and Hr, PSF binds to the DNA-binding domains of NHRs [5].

RIP140

RIP140 represents a third co-repressor category that is able to repress agonist-bound NHRs such as the glucocorticoid receptor or RAR/RXR. Interaction with NHRs is mediated by nine LXXLL receptor-interacting motifs, usually found in co-activators for interaction with NHRs [6,7].

LCoR

In a similar way, LCoR is recruited to agonist-bound NHRs through a single LXXLL motif and represses the activity of oestrogen receptor α, progesterone receptor and vitamin D receptor. Interestingly, LCoR functions by recruitment of HDAC activity and CtBP (C-terminal-binding protein) [8].

Thus, co-repressors exhibit ligand-sensitive, ligand-independent and ligand-dependent interactions with NHRs, and are therefore very important for modulating the activity of NHRs.

NHR action on chromatin

NHRs are able to repress genes through changing chromatin structure and accessibility. The basic structural units of chromatin are the nucleosomes, which consist of DNA wrapped around histone octamers (Figure 3). Histone H1 functions as a linker between the nucleosomes that can lead to an even more compacted organization. Chromatin is not static, instead covalent histone modifications and alterations of the nucleosomal array are two major processes that can influence the accessibility of the transcriptional machinery and thereby the expression of genes.

Histone modifications take place on specific residues most commonly located at the N-terminal tails of the histones and include methylation, phosphorylation and acetylation. It is thought that acetylated histones induce a more open chromatin structure (Figure 3). HDACs remove the acetyl group

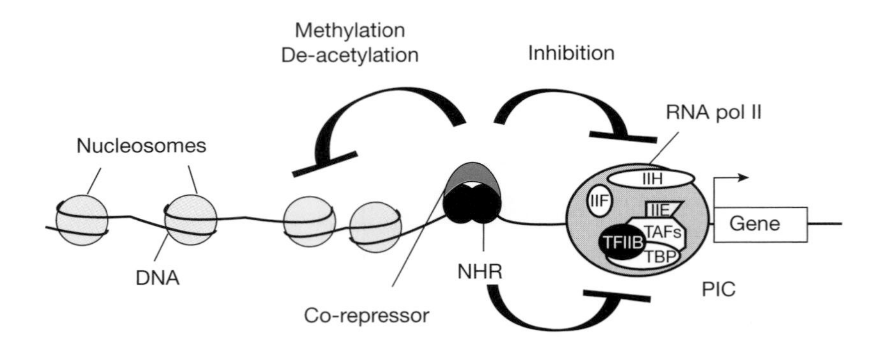

Figure 3. Mechanisms of NHR-mediated gene silencing: chromatin and PIC
Left-hand side: NHRs recruit co-repressors that in turn recruit factors with enzymic activities such as HDACs or histone methyltransferases. Both enzymic activities modify the N-termini of the histones, termed histone tails. HDACs remove acetyl groups that were originally transferred by histone acetyltransferases. Thereby it is thought that histones bind more tightly to DNA and thus lead to a more compact and less accessible chromatin, indicated here as a shorter distance between nucleosomes. Methyltransferases methylate histone tails at lysine and arginine residues. Depending on the specific methylated residue and on how many methyl groups have been transferred, methylation can lead to transcriptional silencing, e.g. through recruitment of heterochromatin-associated protein 1 (HP1; see text). Right-hand side: NHRs interact with factors of the basal transcription machinery of RNA polymerase II (RNA pol II) for both gene activation and gene silencing. Transcription factor IIB (TFIIB) and TATA-binding protein (TBP) interaction with nuclear receptors in the absence of ligand were shown to be associated with gene repression. Furthermore, co-repressors were also shown to interact with the basal transcription machinery. It is thought that the recruitment of basal transcription factors is impaired and/or the activity of the transcriptional PIC (pre-initiation complex) is inhibited. TAF, TBP-associated factor.

from the lysine residues, inducing a more compact structure consistent with findings that repressed genes are mostly hypoacetylated.

Recruitment of HDACs by co-repressors

Co-repressors for NHRs are associated with HDACs [21]. They interact directly with HDAC II family members (HDACs 4–7). In addition, association of co-repressors with class I HDAC complexes have been reported. The Sin3A–Sin-associated protein (SAP) complex harbours the SAPs, the retinoblastoma-associated proteins 46 and 48 (RbAp46 and RbAp48), the HDACs 1 and 2, and the Sin3A protein, which acts as a scaffold for the complex. The class I family member HDAC 3 forms a stable complex with SMRT *in vivo*. Importantly, SMRT and NCoR are activating cofactors for HDAC 3. Activation of HDAC 3 is mediated by the deacetylase-activating domain ('DAD') that includes one of two SANT motifs found in both SMRT and NCoR. The C-terminal SANT motif functions as part of a histone-interaction domain, which enhances repression by increasing the affinity of the deacetylase-activating domain HDAC 3 enzyme for histone substrate and by its ability to inhibit histone acetyltransferase activity [22,23]. So far all co-repressors known for NHRs are associated with HDAC activity, suggesting a major role for gene silencing through histone deacetylation and chromatin compaction.

Recruitment of histone methyltransferases by NHRs

Methylation of the N-terminal histone tails has also been associated with gene repression (Figure 3). This modification is carried out by histone methyltransferases [24]. SUV39H1 (the human variant of suppressor of variegation 3–9) is involved in NHR-mediated repression [25]. It methylates Lys-9 of histone H3 and interacts with unliganded TR, facilitating TR-mediated repression in an histone-methyltransferase-activity-dependent manner. NHRs induce a set of different histone modifications such as acetylation, phosphorylation and methylation. Repression by unliganded TR is accompanied by an increase in His-3–Lys-9 methylation, a decrease in methylation of His-3–Lys-4 and a decrease in phosphorylation of Ser-10 of histone H3. Furthermore, because methylation by SUV39H1 generates a binding motif for heterochromatin-associated protein 1 (HP1), it may promote a higher-order chromatin structure.

Recruitment of nucleosome-remodelling activity by NHR

Also important is the organization and positioning of the nucleosomes. Chromatin-remodelling complexes induce ATP-dependent changes in the position of the histone octamer relative to DNA by a transient disruption of the histone–DNA interaction [26]. Interestingly, the nucleosome remodelling and histone deacetylation (NURD) complex combines HDAC activity and chromatin-remodelling activity. It is proposed that disruption of the nucleosomal array facilitates access of the HDAC component to the histone tails. Studies on TR-dependent repression showed the involvement of the NURD complex [27], which contains the methyl-binding proteins MBD 2 and 3, MTA 2 and the ATP-remodelling component Mi-2 [21]. Both Sin3A and NURD complexes are constitutively associated with chromatin and may thereby be involved in TR-mediated repression by establishing and maintaining a more global histone deacetylation [25].

In addition, NCoR/SMRT form several protein complexes associated with HDAC 3. NCoR, for example, is in a complex containing both HDAC 3 and subunits of a remodelling complex, including Brg1 (brahma-related gene-1), the mammalian Snf2-related ATPase-containing activity and BAF (Brg1-associated factor). The remodelling activity of this complex is also implicated in NHR-mediated repression [28].

Thus NHR-mediated repression can be achieved by recruitment of enzymic complexes that structurally remodel or chemically modify nucleosomes.

Gene silencing by NHRs through interference with the basal transcription machinery

Another mechanism by which NHRs mediate gene silencing is by directly targeting the basal transcriptional machinery (Figure 3). The TR silencing domain interacts with the basal transcription factor TFIIB only in the absence of ligand, suggesting that this interaction contributes to gene silencing [29]. In addition, unliganded TR inhibits the formation of a functional pre-initiation complex (PIC) in *in vitro* chromatin-free assays. In agreement with this, TR inhibits transcription only at an early step during PIC assembly, while pre-assembled PICs are unaffected by TR [30,31]. Using stepwise pre-assembled PICs, it has been shown that TRα targets either the TATA-binding protein (TBP)/TFIIA or the TBP/TFIIA/TFIIB steps of PIC assembly for repression. Furthermore these TR–TBP interactions can be inhibited by thyroid hormone [32].

NCoR/SMRT have also been reported to interact with TFIIB and TBP-associated factors (TAFs) [33,34]. Furthermore, overexpression of NCoR prevents a functional interaction of TFIIB with $TAF_{II}32$, which is a critical step in the assembly of the transcription-initiation complex. This suggests that NCoR also represses gene activity by disturbing the assembly of an intact, functional initiation complex.

Thus, co-repressor activity, as well as NHR-mediated gene silencing, acts on multiple levels, from the basal transcription machinery to chromatin and higher-order structures.

Hormone receptor inactivation by anti-hormones

Anti-hormones are important synthetic pharmaceutical agents that are by definition antagonists of hormone action. They are used in therapies against hormone-dependent tumours, including prostate and breast cancer. Drug targets are the AR in prostate cancer and oestrogen receptor in breast cancer therapy. Anti-hormones bind to the LBD of hormone receptors and inhibit the action of the receptors (Figure 4). In most cases, steroid receptors are inactivated upon anti-hormone binding leading to an inhibition of target-gene transcription, resulting in a block of tumour proliferation [35].

A number of studies suggest that co-repressors are responsible for this effect. An interaction with co-repressors in the presence of anti-hormones has been found for progesterone, glucocorticoid, oestrogen and androgen receptors [36–39] (Figure 4). Crystal structure data show that Helix 12 shifts towards the N-terminus of the antagonist-bound LBD [11]. This shift might not only disrupt co-activator interaction, but might also expose the surface for co-repressor binding. Notably, recent findings for AR suggest that co-repressors are able to bind to the N-terminus of AR to mediate receptor inhibition [37], indicating that co-repressors bind to both C-termini and N-termini of NHRs. These findings provide a molecular explanation for how anti-hormones inhibit steroid receptors: through recruitment of co-repressors.

Negative response elements

NHRs mediate their genomic effects on transcription by binding to specific DNA elements known as hormone-response elements (HREs). Classically,

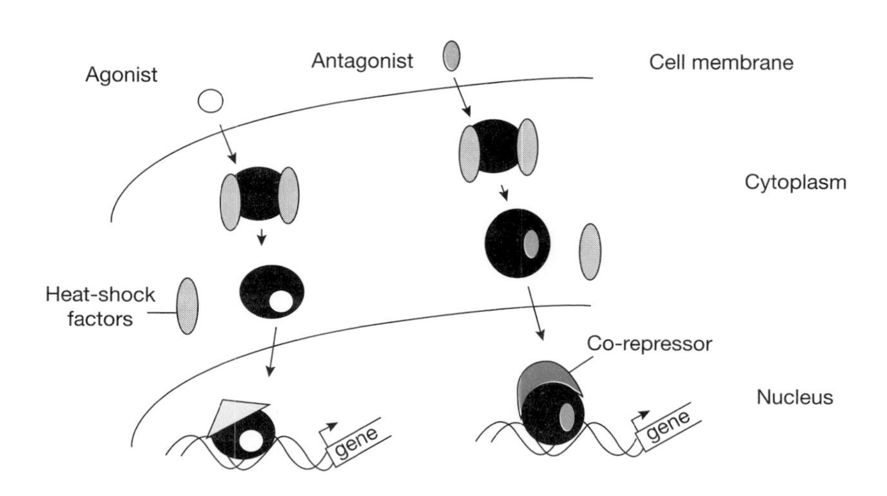

Figure 4. Antagonist-mediated inactivation of receptor and receptor target genes
Inhibition of NHRs by anti-hormones (antagonists) involves the recruitment of co-repressors. Anti-hormones are used for treatment against hormone-regulated growth of cancer, such as breast or prostate cancer. Mechanism of action is exemplified by the AR that shows cytoplasmic localization in the absence of ligand. Ligand induces heat-shock dissociation and nuclear translocation. Agonists induce recruitment of co-activators (triangle) while anti-hormones induce recruitment of co-repressors. Thereby, receptor target genes are inactivated.

hormone binding to receptors leads to hormone-dependent gene activation. However, on response elements known as negative HREs (nHRE), binding of NHRs leads to a ligand-dependent repression of target gene transcription (Figure 5). A common observation is that nHREs are located in the vicinity of the transcriptional start site of a gene or are located downstream of the TATA box [40].

Most of the negative response elements share the common feature that in the absence of hormone type II receptors, such as TR, they bind to these elements to constitutively activate transcription. In contrast, hormone leads to repression of transcription by both class I and class II receptors bound on these negative response elements.

While much work has focused on HREs, only little is known mechanistically about nHREs. On one nHRE it has been shown that the liganded TR recruits HDAC 2 [41]. In addition, it was recently shown that SMRT can function as a co-activator for TRα in the absence of thyroid hormone on another nHRE, suggesting that allosteric changes resulting from binding of TR to different HREs can dictate whether a cofactor acts as a co-activator or co-repressor [42].

Transcriptional repression by NHR through inhibition of signalling pathways via crosstalk

NHRs can also alter expression of genes that do not contain any HREs through interference with other transcription factors (Figure 6). A very well-

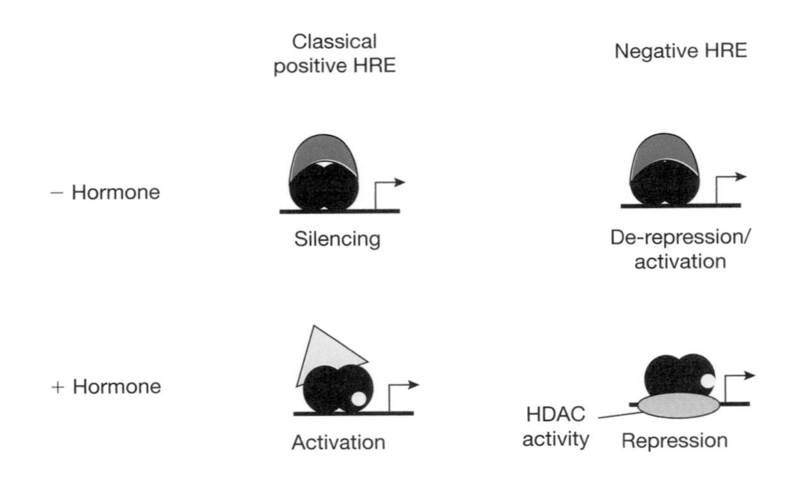

Figure 5. NHRs on negative and positive response elements
In contrast with the recruitment of co-repressors on classical response elements leading to gene silencing, NHRs bound to negative response elements recruit co-repressors for gene activation. In response to hormone (light-blue circle) co-repressors are dissociated and co-activators are recruited on classical response elements, while on negative response elements one report shows that the gene repression is mediated by recruitment of HDACs 1 and 2.

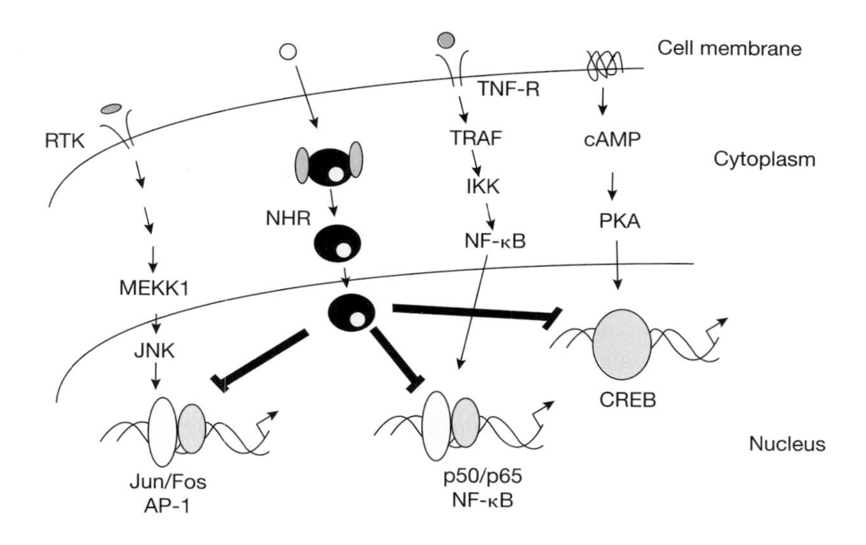

Figure 6. Inhibition of non-NHR signal transduction via crosstalk
NHRs can repress both nuclear factor κB (NF-κB) and Jun/Fos [activator protein 1 (AP-1)]. Both NF-κB and Jun/Fos are activated through signal transducers. Jun/Fos (AP-1) are activated through mitogens and receptor tyrosine kinases (RTK) that transduce the cellular response through various steps to activate c-Jun N-terminal kinase (JNK). Jun and Fos heterodimerize and activate target genes. The NF-κB pathway is activated for example through the tumour necrosis factor receptor (TNF-R), which leads to activation of inhibitory κB (IκB) kinase (IKK), degradation of IκB, translocation of p65/p50 into the nucleus, and gene activation by NF-κB. In addition, liganded NHRs can inactivate cAMP-response-element-binding protein (CREB). NHRs can inhibit these pathways through repression of these DNA-bound transcriptional activators. MEKK1, mitogen-activated protein kinase (MAPK)/extracellular-signal-regulated protein kinase (ERK) kinase kinase 1; TRAF, tumour-necrosis-factor-receptor-associated factor; PKA, protein kinase A.

studied example is the observation that several nuclear receptors, such as TR, RAR or glucocorticoid receptor, can act as ligand-dependent transrepressors of activator protein 1 (AP-1; Jun/Fos) activity, and reciprocally that AP-1 can inhibit transactivation by NHRs [1,40]. It is considered that many anti-proliferative effects of NHR ligands are mediated by their anti-AP-1 activity.

In a similar manner, glucocorticoid receptor can mutually interfere with nuclear factor κB activity (NF-κB), which is a molecular explanation for the anti-inflammatory and immunosuppressive effects of glucocorticoids. Several mechanisms have been proposed for inhibition of AP-1 and NF-κB including direct contact, competition for common transcriptional mediators and involvement of c-Jun N-terminal kinase or other unknown factors [1,40]. Recently, a novel mechanism for inhibition of cAMP-responsive genes by TR was revealed. Here, liganded TR inhibits the transcriptional activation mediated by CREB, the cAMP-response-element-binding protein [43].

Impact of gene repression by NHRs on development and disease

Several observations indicate an important impact of gene repression by NHRs in development and disease. The repression function of TR is implicated in amphibian development. There is evidence for an essential dual role of TR in this process. In premetamorphic tadpoles TR is expressed prior to the production of endogenous thyroid hormone, whereas during metamorphosis production of thyroid hormone allows metamorphosis. Premature exogenous treatment of tadpoles with thyroid hormone leads to premature metamorphosis, suggesting that TR represses metamorphosis in its unliganded form, indicating that TR silences target genes while hormone binding leads to activation of TR target genes involved in metamorphosis [44]. In *Xenopus*, RAR-mediated silencing is critical for head formation. Expression of dominant-negative co-repressors results in malformations similar to early treatment of retinoic acid [45]. NCoR-knockout mice show an altered pattern of transcription, which is associated with developmental blocks at specific points in the central nervous system, erythrocyte and thymocyte differentiation [46].

Aberrant gene silencing is associated with some diseases. An oncogenic form of the TR, v-ErbA, lacking Helix 12, contributes to avian erythroleukaemia by repression of target genes. A point mutant that lacks the ability to repress transcription and to interact with SMRT also abolishes oncogenic transformation. Repression plays also a major role in the syndrome of thyroid hormone resistance, which is caused by mutations in the LBD of TR (see Chapter 12 in this volume). The severity of the resistance directly correlates with the repression function of TR, as mutations of TR found in patients impair the ability of co-repressors to dissociate from the receptor [11,40,47]. In addition, abnormal fusion variants of RARα, created by chromosomal translocations with either the promyelocytic leukaemia or the PLZF (promyelocyte leukaemia zinc finger) locus, block myeloid differentiation, resulting in acute promyelocytic leukaemia. This differentiation block correlates with the ability of the fusion proteins to repress transcription and to recruit nuclear receptor co-repressors. In promyelocytic leukaemia–RARα patients treatment with retinoic acid, which abolishes repression and leads to co-repressor-receptor dissociation, relieves this differentiation block resulting in remission of the disease. However, patients with a PLZF–RAR translocation are not treatable with retinoic acid because there is a second recruitment for co-repressors in the PLZF part that is unaffected by retinoic acid [11,40]. The orphan nuclear receptor DAX-1 (dosage-sensitive hypoplasia congenital on X-chromosome) is implicated in the pathogenesis of adrenal hypoplasia congenita and hypogonadism. DAX-1 inhibits steroidogenic factor 1-mediated transcription and this inhibition is mediated by the C-terminal silencing domain of DAX-1. In most DAX-1 mutants from adrenal hypoplasia congenita patients this region is affected. DAX-1 mutants fail to interact with Alien and have lost their silencing function. This suggests an important role for Alien in the pathogenesis mediated by DAX-1-mutant patients [18,48]. These examples together establish the importance of repression by NHRs in disease and development.

Conclusions

In recent years, it has become more and more evident that transcriptional repression is as important as transcriptional activation in the control of gene expression. In this chapter, we have reviewed the different levels and mechanisms of repression by NHRs. Genomic effects are mediated by recruitment of co-repressors, effects on chromatin structure, contacts to the basal transcription machinery and negative response elements. In addition, repression is achieved via crosstalk with other signalling pathways, like AP-1. Thus, repression by NHRs is a complex, combinatorial, multilevel process. We are still at the beginning of understanding the complex interplay of activation and repression. Microarrays will be important tools to identify target genes silenced by NHR. Future experiments including mice model systems will reveal the biological role of the different levels of repression.

Summary

- *Gene silencing by NHRs is essential for proper development.*
- *Aberrant gene silencing leads to disease, cancer and malformations.*
- *Gene silencing is mediated by inhibition of the basal transcription apparatus.*
- *Gene silencing is also mediated at the chromatin level by histone deacetylation, methylation and alteration of the nucleosomal array.*
- *NHRs repress other signalling pathways through crosstalk.*
- *NHR repression is a major drug target for cancer therapy (hormone antagonists and agonists, such as retinoic acid).*

Due to space limitations we apologize to all authors in this field for not citing their contribution. We thank Dr Leslie Burke for critical reading of the manuscript and helpful discussions.

References

1. Gottlicher, M., Heck, S. & Herrlich, P. (1998) Transcriptional cross-talk, the second mode of steroid hormone receptor action. *J. Mol. Med.* **76**, 480–489
2. Mangelsdorf, D.J., Thummel, C., Beato, M., Herrlich, P., Schutz, G., Umesono, K., Blumberg, B., Kastner, P., Mark, M., Chambon, P. et al. (1995) The nuclear receptor superfamily: the second decade. *Cell* **83**, 835–839
3. Burke, L.J. & Baniahmad, A. (2000) Co-repressors 2000. *FASEB J.* **14**, 1876–1888
4. Zamir, I., Dawson, J., Lavinsky, R.M., Glass, C.K., Rosenfeld, M.G. & Lazar, M.A. (1997) Cloning and characterization of a corepressor and potential component of the nuclear hormone receptor repression complex. *Proc. Natl. Acad. Sci. U.S.A.* **94**, 14400–14405

5. Mathur, M., Tucker, P.W. & Samuels, H.H. (2001) PSF is a novel corepressor that mediates its effect through Sin3A and the DNA binding domain of nuclear hormone receptors. *Mol. Cell. Biol.* **21**, 2298–2311
6. Zilliacus, J., Holter, E., Wakui, H., Tazawa, H., Treuter, E. & Gustafsson, J.-Å. (2001) Regulation of glucocorticoid receptor activity by 14-3-3-dependent intracellular relocalization of the corepressor RIP140. *Mol. Endocrinol.* **15**, 501–511
7. Lee, C.H. & Wei, L.N. (1999) Characterization of receptor-interacting protein 140 in retinoid receptor activities. *J. Biol. Chem.* **274**, 31320–31326
8. Fernandes, I., Bastien, Y., Wai, T., Nygard, K., Lin, R., Cormier, O., Lee, H.S., Eng, F., Bertos, N.R., Pelletier, N. et al. (2003) Ligand-dependent nuclear receptor corepressor LCoR functions by histone deacetylase-dependent and -independent mechanisms. *Mol. Cell* **11**, 139–150
9. Rosenfeld, M.G. & Glass, C.K. (2001) Coregulator codes of transcriptional regulation by nuclear receptors. *J. Biol. Chem.* **276**, 36865–36868
10. Nagy, L., Kao, H.Y., Love, J.D., Li, C., Banayo, E., Gooch, J.T., Krishna, V., Chatterjee, K., Evans, R.M. & Schwabe, J.W. (1999) Mechanism of corepressor binding and release from nuclear hormone receptors. *Genes Dev.* **13**, 3209–3216
11. Hu, X. & Lazar, M.A. (2000) Transcriptional repression by nuclear hormone receptors. *Trends Endocrinol. Metab.* **11**, 6–10
12. Ishizuka, T. & Lazar, M.A. (2003) The N-CoR/histone deacetylase 3 complex is required for repression by thyroid hormone receptor. *Mol. Cell. Biol.* **23**, 5122–5131
13. Lin, B.C., Hong, S.H., Krig, S., Yoh, S.M. & Privalsky, M.L. (1997) A conformational switch in nuclear hormone receptors is involved in coupling hormone binding to corepressor release. *Mol. Cell. Biol.* **17**, 6131–6138
14. Baniahmad, A., Dressel, U. & Renkawitz, R. (1998) Cell-specific inhibition of retinoic acid receptor-alpha silencing by the AF2/tau c activation domain can be overcome by the corepressor SMRT, but not by N-CoR. *Mol. Endocrinol.* **12**, 504–512
15. Tsai, C.C., Kao, H.Y., Yao, T.P., McKeown, M. & Evans, R.M. (1999) SMRTER, a *Drosophila* nuclear receptor coregulator, reveals that EcR- mediated repression is critical for development. *Mol. Cell* 4, 175–186
16. Dressel, U., Thormeyer, D., Altincicek, B., Paululat, A., Eggert, M., Schneider, S., Tenbaum, S.P., Renkawitz, R. & Baniahmad, A. (1999) Alien, a highly conserved protein with characteristics of a corepressor for members of the nuclear hormone receptor superfamily. *Mol. Cell. Biol.* **19**, 3383–3394
17. Polly, P., Herdick, M., Moehren, U., Baniahmad, A., Heinzel, T. & Carlberg, C. (2000) VDR-Alien: a novel, DNA-selective vitamin D(3) receptor-corepressor partnership. *FASEB J.* **14**, 1455–1463
18. Altincicek, B., Tenbaum, S.P., Dressel, U., Thormeyer, D., Renkawitz, R. & Baniahmad, A. (2000) Interaction of the corepressor Alien with DAX-1 is abrogated by mutations of DAX-1 involved in adrenal hypoplasia congenita. *J. Biol. Chem.* **275**, 7662–7667
19. Tenbaum, S.P., Juenemann, S., Schlitt, T., Bernal, J., Renkawitz, R., Munoz, A. & Baniahmad, A. (2003) Alien/CSN2 gene expression is regulated by thyroid hormone in rat brain. *Dev. Biol.* **254**, 149–160
20. Potter, G.B., Beaudoin, 3rd, G.M., DeRenzo, C.L., Zarach, J.M., Chen, S.H. & Thompson, C.C. (2001) The hairless gene mutated in congenital hair loss disorders encodes a novel nuclear receptor corepressor. *Genes Dev.* **15**, 2687–2701
21. Jepsen, K. & Rosenfeld, M.G. (2002) Biological roles and mechanistic actions of co-repressor complexes. *J. Cell Sci.* **115**, 689–698
22. Guenther, M.G., Barak, O. & Lazar, M.A. (2001) The SMRT and N-CoR corepressors are activating cofactors for histone deacetylase 3. *Mol. Cell. Biol.* **21**, 6091–6101
23. Yu, J., Li, Y., Ishizuka, T., Guenther, M.G. & Lazar, M.A. (2003) A SANT motif in the SMRT corepressor interprets the histone code and promotes histone deacetylation. *EMBO J.* **22**, 3403–3410
24. Kouzarides, T. (2002) Histone methylation in transcriptional control. *Curr. Opin. Genet. Dev.* **12**, 198–209

25. Li, J., Lin, Q., Yoon, H.G., Huang, Z.Q., Strahl, B.D., Allis, C.D. & Wong, J. (2002) Involvement of histone methylation and phosphorylation in regulation of transcription by thyroid hormone receptor. *Mol. Cell. Biol.* **22**, 5688–5697
26. Becker, P. & Hoerz, W. (2002) ATP-dependent nucleosome remodeling. *Annu. Rev. Biochem.* **71**, 247–273
27. Xue, Y., Wong, J., Moreno, G.T., Young, M.K., Cote, J. & Wang, W. (1998) NURD, a novel complex with both ATP-dependent chromatin-remodeling and histone deacetylase activities. *Mol. Cell* **2**, 851–861
28. Jung, D.J., Lee, S.K. & Lee, J.W. (2001) Agonist-dependent repression mediated by mutant estrogen receptor α that lacks the activation function 2 core domain. *J. Biol. Chem.* **276**, 37280–37283
29. Baniahmad, A., Ha, I., Reinberg, D., Tsai, S., Tsai, M.J. & O'Malley, B.W. (1993) Interaction of human thyroid hormone receptor β with transcription factor TFIIB may mediate target gene derepression and activation by thyroid hormone. *Proc. Natl. Acad. Sci. U.S.A.* **90**, 8832–8836
30. Fondell, J.D., Roy, A.L. & Roeder, R.G. (1993) Unliganded thyroid hormone receptor inhibits formation of a functional preinitiation complex: implications for active repression. *Genes Dev.* **7**, 1400–1410
31. Tong, G.X., Tanen, M.R. & Bagchi, M.K. (1995) Ligand modulates the interaction of thyroid hormone receptor β with the basal transcription machinery. *J. Biol. Chem.* **270**, 10601–10611
32. Fondell, J.D., Brunel, F., Hisatake, K. & Roeder, R.G. (1996) Unliganded thyroid hormone receptor α can target TATA-binding protein for transcriptional repression. *Mol. Cell. Biol.* **16**, 281–287
33. Muscat, G.E., Burke, L.J. & Downes, M. (1998) The corepressor N-CoR and its variants RIP13a and RIP13Δ1 directly interact with the basal transcription factors TFIIB, TAFII32 and TAFII70. *Nucleic Acids Res.* **26**, 2899–2907
34. Wong, C.W. & Privalsky, M.L. (1998) Transcriptional repression by the SMRT-mSin3 corepressor: multiple interactions, multiple mechanisms, and a potential role for TFIIB. *Mol. Cell. Biol.* **18**, 5500–5510
35. Fuhrmann, U., Parczyk, K., Klotzbucher, M., Klocker, H. & Cato, A.C. (1998) Recent developments in molecular action of antihormones. *J. Mol. Med.* **76**, 512–524
36. Schulz, M., Eggert, M., Baniahmad, A., Dostert, A., Heinzel, T. & Renkawitz, R. (2002) RU486-induced glucocorticoid receptor agonism is controlled by the receptor N terminus and by corepressor binding. *J. Biol. Chem.* **277**, 26238–26243
37. Dotzlaw, H., Moehren, U., Mink, S., Cato, A.C., Iniguez Lluhi, J.A. & Baniahmad, A. (2002) The amino terminus of the human AR is target for corepressor action and antihormone agonism. *Mol. Endocrinol.* **16**, 661–673
38. Jackson, T.A., Richer, J.K., Bain, D.L., Takimoto, G.S., Tung, L. & Horwitz, K.B. (1997) The partial agonist activity of antagonist-occupied steroid receptors is controlled by a novel hinge domain-binding coactivator L7/SPA and the corepressors N-CoR or SMRT. *Mol. Endocrinol.* **11**, 693–705
39. Smith, C.L., Nawaz, Z. & O'Malley, B.W. (1997) Coactivator and corepressor regulation of the agonist/antagonist activity of the mixed antiestrogen, 4-hydroxytamoxifen. *Mol. Endocrinol.* **11**, 657–666
40. Aranda, A. & Pascual, A. (2001) Nuclear hormone receptors and gene expression. *Physiol. Rev.* **81**, 1269–1304
41. Sasaki, S., Lesoon-Wood, L.A., Dey, A., Kuwata, T., Weintraub, B.D., Humphrey, G., Yang, W.M., Seto, E., Yen, P.M., Howard, B.H. & Ozato, K. (1999) Ligand-induced recruitment of a histone deacetylase in the negative-feedback regulation of the thyrotropin β gene. *EMBO J.* **18**, 5389–5398
42. Berghagen, H., Ragnhildstveit, E., Krogsrud, K., Thuestad, G., Apriletti, J. & Saatcioglu, F. (2002) Corepressor SMRT functions as a coactivator for thyroid hormone receptor T3Rα from a negative hormone response element. *J. Biol. Chem.* **277**, 49517–49522
43. Mendez-Pertuz, M., Sanchez-Pacheco, A. & Aranda, A. (2003) The thyroid hormone receptor antagonizes CREB-mediated transcription. *EMBO J.* **22**, 3102–3112

44. Sachs, L.M. & Shi, Y.B. (2000) Targeted chromatin binding and histone acetylation in vivo by thyroid hormone receptor during amphibian development. *Proc. Natl. Acad. Sci. U.S.A.* **97**, 13138–13143
45. Koide, T., Downes, M., Chandraratna, R.A., Blumberg, B. & Umesono, K. (2001) Active repression of RAR signaling is required for head formation. *Genes Dev.* **15**, 2111–2121
46. Jepsen, K., Hermanson, O., Onami, T.M., Gleiberman, A.S., Lunyak, V., McEvilly, R.J., Kurokawa, R., Kumar, V., Liu, F., Seto, E. et al. (2000) Combinatorial roles of the nuclear receptor corepressor in transcription and development. *Cell* **102**, 753–763
47. Yoh, S.M., Chatterjee, V.K. & Privalsky, M.L. (1997) Thyroid hormone resistance syndrome manifests as an aberrant interaction between mutant T3 receptors and transcriptional corepressors. *Mol. Endocrinol.* **11**, 470–480
48. Tenbaum, S. & Baniahmad, A. (1997) Nuclear receptors: structure, function and involvement in disease. *Int. J. Biochem. Cell Biol.* **29**, 1325–1341

8

Receptor mechanisms of rapid extranuclear signalling initiated by steroid hormones

Viroj Boonyaratanakornkit and Dean P. Edwards[1]

Department of Pathology, School of Medicine and Program in Molecular Biology, University of Colorado Health Sciences Center, Denver, CO 80218, U.S.A.

Abstract

In addition to their role as direct regulators of gene transcription mediated by classical nuclear hormone receptors, steroid hormones have also been described to exert rapid effects on intracellular signalling pathways independent of gene transcription. This chapter focuses on recent advances in our understanding of the receptors and mechanisms that mediate these rapid signalling actions of oestrogens and progesterone. Increasing evidence suggests that at least some of these rapid actions are mediated by a subpopulation of the classical nuclear oestrogen receptor (ER) and progesterone receptor (PR) that localize to the cytoplasm or associate with the plasma membrane. Human PR has been shown to mediate rapid progestin activation of the Src/Ras/Raf/mitogen-activated protein kinase signalling pathway in mammalian cells by a direct interaction with the Src homology 3 domain of Src tyrosine kinases through a Pro-Xaa-Xaa-Pro-Xaa-Arg motif located in the N-terminal domain of the receptor. Moreover, this is an extranuclear action of PR that is separable from its direct transcriptional activity. Additionally, a novel membrane protein unrelated to nuclear PR was recently identified that has

[1]*To whom correspondence should be addressed (e-mail dean.edwards@uchsc.edu).*

properties of a G-protein-coupled receptor for progesterone and has been shown to be involved in mediating the extranuclear signalling actions of progesterone that promotes oocyte maturation in fish. The role of this membrane PR (mPR) in mammalian cells is less clear and the relationship of the membrane and classical nuclear PR in mediating rapid non-transcriptional signalling of progestins has not been explored. To date, a novel membrane ER unrelated to classical nuclear receptors has not been cloned and characterized, and many of the known rapid extranuclear signalling actions of oestrogen appear also to be mediated by a subpopulation of nuclear ER, or a closely related receptor. A novel protein termed modulator of non-genomic activity of ER (MNAR) has been identified that acts as an adaptor between ER and Src, and thus provides a mechanisms for coupling of oestrogen and ER with rapid oestrogen-induced activation of Src and the downstream mitogen-activated protein kinase signalling cascade. The physiological relevance of rapid extranuclear signalling by the classical ER has been provided by experiments showing that these actions contribute to the anti-apoptotic effect of oestrogen in bone *in vivo* and to the rapid effects of oestrogen on vasodilation and protection of endothelial cells against injury.

Introduction

Steroid hormones exert many of their biological responses through nuclear receptors that directly regulate specific networks of gene expression. Nuclear receptors belong to a superfamily of ligand-dependent transcription factors that share a similar domain organization consisting of a centrally located, highly conserved DNA-binding domain, a C-terminal ligand-binding domain (LBD) and an N-terminal domain that is required for full transcription activity (Figure 1). Hormone binding activates the transcriptional potential of steroid hormone receptors by inducing a conformational change leading to dissociation from molecular chaperones, dimerization and binding to specific DNA sequences in the regulatory promoter regions of steroid-responsive genes, referred to as hormone-response elements. Upon binding to DNA, activated receptors recruit specific co-activators that are essential for assembly of a productive transcriptional complex.

However, not all biological effects of steroid hormones can be explained by the well-established roles of steroid hormone receptors as transcription factors. All classes of steroid hormone have been observed to have rapid effects that occur independently of gene transcription. This phenomenon has been termed 'non-genomic' action of steroids to distinguish it from the classical direct genomic actions of steroid receptors in the nucleus. These rapid actions of steroids have been described to generate intracellular second messengers, activate or inhibit cell signal-transduction cascades and modulate ion channels. However, because some of the intracellular signalling pathways activated by steroids can converge upon and activate other nuclear transcription factors, the

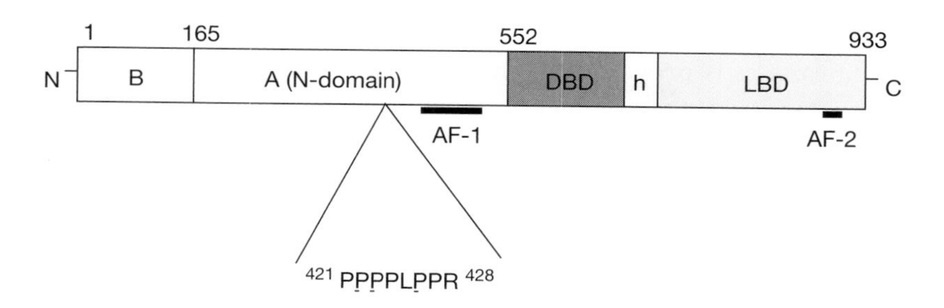

Figure 1. Schematic of PR domains and the PXXPXR motif
DBD, DNA-binding domain; A, N-terminal truncated; PR-A; B, full length PR-B; h, hinge region.

term 'non-genomic' is not very accurate. We propose the more appropiate terminology 'extranuclear steroid-initiated signalling'.

The receptors and mechanisms that mediate extranuclear signalling by steroid hormones are generally not well defined and these are intensive on-going areas of investigation. Studies have indicated in part that a subpopulation of classical steroid receptors can mediate extranuclear signalling actions of steroids by somehow associating with the cell membrane and/or signalling complexes in the cytoplasm. Other studies suggest the existence of separate membrane receptors unrelated to the classical steroid receptors. All classes of steroid hormone have been described to exert extranuclear actions on cell signal-transduction cascades and this has recently been reviewed in detail elsewhere [1]. The focus of this review is on recent studies defining the receptors and mechanisms of rapid extranuclear signalling by oestrogen and progesterone.

Extranuclear signalling actions of progesterone: classical intracellular progesterone receptor (PR)

The PR is expressed in most target tissues as two isoforms, PR-A and PR-B, that arise from alternative uses of two promoters from the same gene. In human cells PR-A is truncated, lacking the first 164 amino acids of the N-terminal domain of full-length PR-B (Figure 1). The transcriptional activities of the two PR isoforms vary depending on the cell type and promoter context. In general, PR-B is a stronger transcriptional activator whereas PR-A can function as a ligand-dependent repressor of other steroid hormone receptors including PR-B, oestrogen receptor (ER), androgen receptor (AR), glucocorticoid receptor and mineralocorticoid receptor [2].

Rapid, extranuclear signalling actions of progesterone have been described in several biological systems. These include progesterone-induced maturation of the *Xenopus* oocyte, stimulation of acrosome reaction in sperm, modulation of neurotransmitter and neuron excitability, and activation of the Src/Ras/mitogen-activated protein kinase [MAPK; extracellular-signal-regulated kinase (ERK)] signal transduction pathway in breast cancer and other mammalian cells.

Xenopus oocyte maturation: role of classical PR

The most thoroughly studied system of rapid extranuclear signalling by steroid hormones is progesterone induction of *Xenopus* oocyte maturation. Progesterone treatment of *Xenopus* oocytes causes a rapid decrease in cAMP and a decline in protein kinase A activity that leads to activation of a MAPK pathway and maturation-promoting factor, causing germinal vesicle breakdown and resumption of meiosis II [3]. A *Xenopus* homologue of the classical mammalian nuclear PR (XPR) was recently cloned and characterized, but shows a somewhat different cellular localization from mammalian nuclear PR [4,5]. XPR localizes predominantly to the cytoplasm [5] with small amounts in the cell membrane [7], and little or no nuclear localization [5] in *Xenopus* oocytes. However, in mammalian cells (COS cells) transiently transfected with XPR cDNA, XPR localized mainly to the nucleus and exhibited progesterone-induced transcriptional activation of a PRE reporter gene [5]. Injection of XPR mRNA increased the sensitivity of *Xenopus* oocyte to progesterone and accelerated progesterone-induced p42 MAPK activation and oocyte maturation. Injection of XPR antisense oligonucleotide inhibited progesterone-induced oocyte maturation while the transcription inhibitor, actinomycin D, did not alter the influence of XPR on oocyte maturation. In addition to XPR, injection of classical human PR has also been shown to accelerate progesterone-induced *Xenopus* oocyte maturation [6]. More recently, XPR was shown to associate with active phosphoinositide 3-kinase (PI 3-kinase) and p42 MAPK upon progesterone treatment, suggesting a direct association of XPR with cytoplasmic signalling pathways [7]. However, progesterone-induced oocyte maturation still occurs when XPR is inhibited, suggesting that XPR is involved but is not sufficient for progesterone-induced oocyte maturation. Taken together, these data suggest that progesterone-induced oocyte maturation is mediated in part by the classical PR and that a separate unrelated mPR may also be involved.

Human PR mediates rapid progesterone induction of Src/Ras/Raf/MAPK signalling in mammalian cells

Rapid progestin activation of the Src/Ras/MAPK pathway in the absence of transcription was first shown by Migiliaccio et al. [8] to be dependent on the classical nuclear PR. Furthermore, the proliferative effect of progesterone in breast cancer cells was shown to be mediated in part by the Src/Ras/MAPK pathway. Studies by Migiliaccio et al. [8] further suggested that PR activation of Src/Ras/MAPK pathway was indirect through PR association with ER, and that it was ER that directly activated Src through interaction with the Src homology 2 domain (SH2 domain). However, whether PR and ER can physically interact is controversial, and how ER interaction can activate Src has not been shown [8]. More recently, an alternative mechanism for how PR interacts with and activates Src and the downstream MAPK cascade was reported [6].

The N-terminal region common to human PR-A and PR-B was discovered to contain a short, contiguous polyproline sequence (amino acids 421–428, Pro-Pro-Pro-Pro-Leu-Pro-Pro-Arg; Figure 1) that conforms to a consensus type II, Pro-Xaa-Xaa-Pro-Xaa-Arg (PXXPXR), motif for binding the SH3 domain of cytoplasmic signalling molecules. These polyproline sequences form a left-handed helix that interacts with the binding pocket of SH3 domains. PR, through this PXXPXR motif, interacts directly *in vitro* with the SH3 domain of various signalling molecules, including Src. PR interaction with the SH3 domain of Src through the PXXPXR motif also occurs in cells in a progesterone-dependent manner. The consequence of this interaction, both *in vitro* (cell-free) and in intact cells, is a potent activation of Src catalytic activity through an SH3-domain-displacement mechanism. The Src-family kinases are autoinhibited by an intracellular association between the SH2 domain and a C-terminal tyrosine phosphorylation site, and between the SH3 domain and a polyproline-like motif in the linker region. These intramolecular interactions maintain the kinase in an inactive closed conformation (Figure 2A) [9]. The PXXPXR motif in PR acts as an external peptide ligand to disrupt the intramolecular SH3 domain interaction and convert Src into an active open conformation (Figure 2A). The PR concentration that gives half-maximal activation of Src kinase is in the low nanomolar range, indicating that PR is a potent activator of Src-family kinases [6].

PR interaction with SH3 domains and activation of Src appears to be of biological consequence in the cell. Progesterone-induced activation of the entire Src/Ras/MAPK signalling pathway in mammalian cells is dependent on the integrity of PXXPXR motif in PR and the ability of PR to interact with the SH3 domain of Src. Thus PR activation of Src is not spurious and is sufficient to generate activation of the entire MAPK phosphorylation cascade. In addition, the PR–SH3-domain interaction contributes to progesterone-induced inhibition of cell proliferation and cell-cycle progression of normal mammary epithelial cells in culture. Interestingly, point mutations in the PXXPXR motif that abolish progesterone-induced Src activation do not influence the transcriptional activity of PR. Conversely, point mutations in the DNA-binding domain or AF-2 (activation function 2) that compromise PR function as a transcription factor have no effect on the ability of PR to mediate progesterone-induced activation of Src [6]. These experimental data indicate that transcription and non-transcription actions of PR are separable. Taken together, these data suggest that PR is a dual-function protein capable of directly interacting with DNA in the nucleus and functioning in its well-established role as a transcription factor, and interacting with SH3 domains of Src and perhaps other signalling molecules to modulate intracellular signalling pathways (Figure 3).

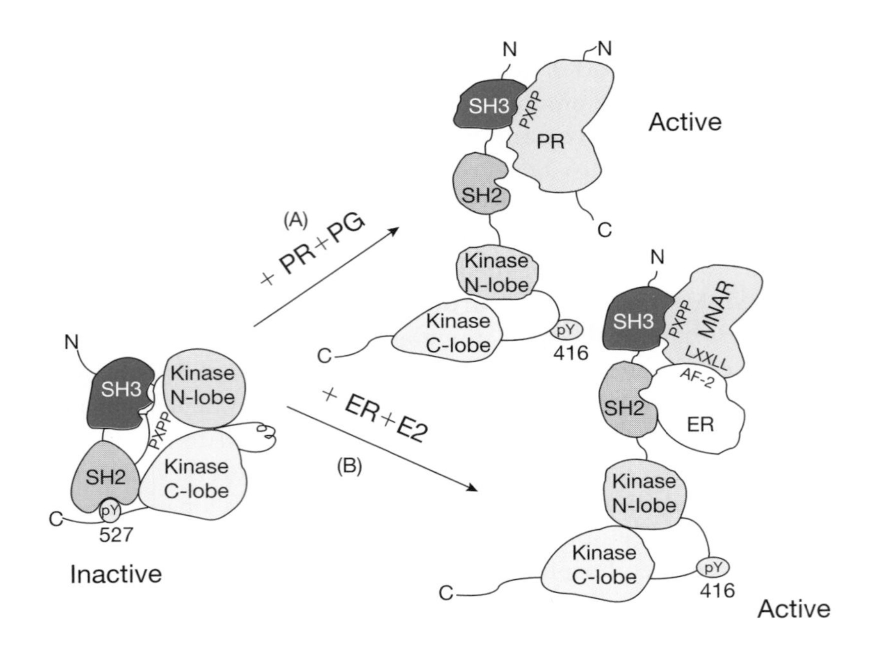

Figure 2. Proposed mechanisms for PR and ER activation of Src kinases
(**A**) The PXXPXR motif in the N-terminal domain of PR interacts with the SH3 domain of Src and converts Src from an inactive 'closed conformation' into an active 'open conformation' by an SH3-displacement mechanism. (**B**) ER in the presence of oestrogen (E2) interacts with MNAR (modulator of non-genomic activity of ER) and Src to form a stable ternary complex through a specific oestrogen-induced interaction between the AF-2 domain of ER and LXXLL motifs in MNAR. Additional stabilizing interactions are provided by ER interaction (possibly through the ER LBD) with the SH2 domain of Src. The PXXPXR motif in MNAR interacts with the SH3 domain of Src and activates Src by an SH3-displacement mechanism. PG, progesterone; pY, phosphotyrosine.

Extranuclear progesterone signalling: role of a novel mPR

Some extranuclear signalling actions of progesterone have long been postulated to be mediated by a distinct membrane receptor, unrelated to classical intracellular PR. In particular, data on *Xenopus* oocyte maturation indicate the involvement of G-proteins and adenylate cyclase, and the response exhibits a steroid specificity that cannot be explained entirely by classical PR. A novel G-protein-coupled receptor (GPCR) was recently cloned and characterized from sea trout ovaries by a receptor-capture assay. A partially purified putative membrane receptor was used to generate monoclonal antibodies which in turn were used to screen a sea trout ovarian expression cDNA library [10]. There is strong evidence that the isolated clone encodes a bona fide mPR that is structurally unrelated to the classical nuclear PR. Recombinant mPR expressed in *Escherichia coli* binds specifically to progestins *in vitro* with a relatively high affinity (K_d, 30 nM), consistent with physiological concentrations of progesterone in sea trout ovaries. Deduced amino acid sequences from mPR cDNA predicted seven transmembrane domains, a characteristic of GPCRs,

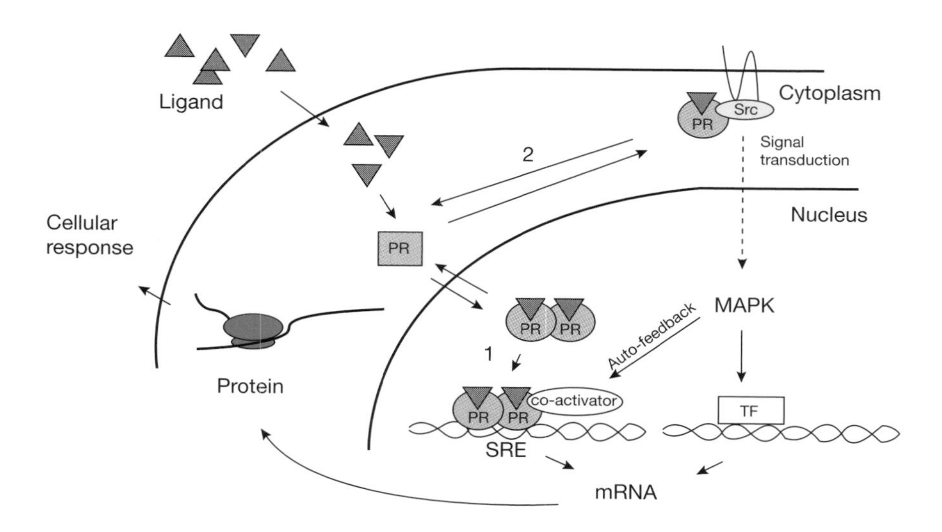

Figure 3. Model of direct transcription and extranuclear signalling pathways of PR
(1) Nuclear transcription pathways. Progestins activate PR by inducing a conformational change(s) that leads to nuclear translocation, dimerization and binding to steroid-response elements (SRE; or hormone-response elements). Activated PR recruits co-activators that are essential for assembly of a productive transcription complex at the promoter. (2) Extranuclear signalling pathway. A subpopulation of PR associates in a progestin-dependent manner with cytoplasmic and/or cell-membrane-signalling molecules including the tyrosine kinase Src. PR–Src interaction is responsible for mediating progesterone-induced activation of the Src/Raf/MAPK phosphorylation cascade. Because MAPK can directly or indirectly activate other transcription factors (TF), progesterone activation of MAPK can potentially regulate sets of genes that are distinct from or complementary to those regulated by direct nuclear transcriptional pathways. Progestin-activated MAPK can also enhance the direct transcriptional activity of PR by an auto-feedback loop through phosphorylation of PR or PR-associated co-activators.

but not similar to nuclear PR including the LBD. Moreover, mPR-mediated progesterone inhibition of adenylate cyclase can be inhibited by pertussis toxin treatment, consistent with coupling of the receptor to a G-protein, more specifically to $G_{i/o}$ proteins [10]. Progestin and gonadotropin increase both mRNA and protein levels of mPR in sea trout ovaries, further implicating mPR as a physiological mediator of progesterone-induced oocyte maturation. Furthermore, injection of mPR antisense oligonucleotides blocked progesterone-mediated oocyte maturation [10].

There are three isoforms of the membrane receptor, which are relatively well conserved from sea trout to human (α, β and γ) [11]. Of interest are the α and β isoforms. In humans, the α isoform is expressed mainly in reproductive tissues, suggesting a role in extranuclear progesterone signalling in oocyte and sperm, whereas the β isoform is mainly expressed in the brain. The γ isoform is mainly expressed in kidney, lung and colon. However, the biological roles and the structure–function relationships of these novel mPRs have yet to be determined. The mPR so far has been studied most extensively for its role in mediating progesterone-induced maturation of ovaries in non-mammalian systems.

However, mPR may also have a role in mammalian cells. In human breast cancer cells (MDA-MB-231), which do not express classical nuclear PR, a rapid and transient progesterone-mediated MAPK activation and inhibition of adenylate cyclase was observed when cells were stably transfected with mPR cDNA [10].

It is not clear whether the classical PR, like mPR, can physically couple with and mediate its effect through G-proteins. Recent evidence suggested that an orphan GPCR, GPR30, could play a role in progesterone-induced growth inhibition of human breast cancer cells (MCF-7 cells). Overexpression of GPR30 inhibited growth of MCF-7 cells, while antisense inhibition of GPR30 expression reversed the growth-inhibitory effect of progesterone [12]. Whether or how GPR30 interacts with the classical PR is not known.

Extranuclear signalling of oestrogen: role of the classical ER

Classical intracellular ER exists as two isoforms synthesized from separate genes, ERα and ERβ. Selective deletion of ERα or ERβ in mice suggests that the effects of oestrogen on development and function of female reproductive tissues are mediated predominantly by ERα, while ERβ is important in the development of the ovary [13]. Both isoforms of ER are also expressed in non-reproductive tissue as well, including bone, cardiovascular tissue, central nervous system and brain. In addition to its well-established biological roles in reproductive tissue, oestrogen also has important biological roles in non-reproductive tissue including maintenance of bone density, vascular wall tone and protection against vascular injury and specific neural functions.

Oestrogen induces rapid activation of various signalling molecules and signal transduction cascades within seconds or minutes, and in a manner unaffected by RNA and protein-synthesis inhibitors. These extranuclear signalling actions of oestrogen are cell- and tissue-specific, and in many systems appear to be mediated by a subpopulation of classical ERα or ERβ. In breast cancer cells, oestrogen induces a rapid activation of the Src/Shc/Ras/Raf/MAPK [14–16] and PI 3-kinase/Akt [17] pathways in an ERα-dependent manner. Both signalling pathways have been shown to be involved in mediating the proliferative response of oestrogens. In osteoblasts and osteocytes, the anti-apoptotic effect of oestrogen is mediated by rapid ER activation of the Src/MAPK pathway [18]. This rapid extranuclear signalling by oestrogen appears to contribute significantly to the maintenance of bone density. In endothelial cells, ERα, by a mechanism that involves association with the p85 subunit of the PI 3-kinase and activation of the PI 3-kinase/Akt signalling pathway, mediates rapid oestrogen activation of the endothelial form of nitric oxide synthase (eNOS) [19]. In addition, ERα mediated rapid oestrogen activation of the p38-β isoform of MAPK in endothelial cells and has been implicated in the protective effect of oestrogen against endothelial cell injury. Activation of c-Jun N-terminal kinase by ER has also been shown to be involved in mediating the anti-apoptotic effect of oestrogen in breast cancer cells [20].

Although an oestrogen-dependent interaction of ER with Src [6,17], Shc [16] and the p85 subunit of PI 3-kinase [19] has been reported, how ER interacts with these molecules in an oestrogen-dependent manner and how these interactions trigger a signalling response remains unclear. Recently, Wong et al. [21] identified a novel protein termed modulator of non-genomic activity of ER (MNAR) that acts as an adapter to couple ER with Src. MNAR facilitates and stabilizes oestrogen-dependent ER–Src interaction by forming a ternary ER–MNAR–Src complex. MNAR also enhances ER-dependent oestrogen activation of Src activity both in cell-free extracts and in intact cells. An interesting feature of MNAR is the presence of both LXXLL and PXXPXR motifs in the same molecule, suggesting a mechanism for how MNAR functions as an adapter between ER and Src. The AF-2 domain of ER interacts in a ligand-dependent manner with LXXLL motifs present in the p160 family of steroid hormone receptor co-activators. Interestingly, oestrogen antagonists induce an alternative conformation in AF-2 that does not permit this interaction [22], suggesting a mechanism for how ER antagonists block rapid oestrogen responses on Src-initiated signalling pathways. The PXXPXR motif present in MNAR is proposed to interact with the SH3 domain of Src and to activate Src through an SH3-domain-displacement mechanism similar to the direct activation by PR. A proposed mechanism of how MNAR mediates ER activation of Src is shown in Figure 2(B). The LXXLL motifs of MNAR mediate oestrogen-dependent interaction with AF-2 in the LBD of ER [21,23], while the PXXPXR motif in MNAR interacts with the SH3 domain of Src to activate Src by an SH3-domain-displacement mechanism. Interaction of ER with the SH2 domain of Src may provide additional stabilizing interactions to contribute to Src activation.

In cell-transfection assays, MNAR potentiated the direct transcriptional activity of ER but this effect was blocked by inhibitors of Src and MAPKs [21]. These inhibitors do not affect the ER co-activation function of p160 co-activators; thus it is not likely that MNAR functions as a conventional ER co-activator. Potentiation of the transcriptional activity of ER is probably due to positive feedback between the oestrogen-stimulated Src/ERK signalling pathway and the ER transcription complex. Previous studies have shown that activation of the MAPK signalling pathway can stimulate the transcriptional activity of ER by phosphorylation and activation of AF-1. Therefore MNAR may also provide a link between transcriptional and non-transcriptional actions of ER.

Extranuclear signalling actions of oestrogen: role of novel membrane receptors?

The existence of a membrane ER, unrelated to classical intracellular ER, has also been long postulated. However, the cloning and characterization of such a novel membrane ER analogous to the recently identified mPR has not yet been accomplished. An orphan GPCR, GPR30, has been implicated in rapid oestrogen signalling activation of MAPK in breast cancer cells by an undefined mechanism [23a]. Whether GPR30 can directly bind oestrogen and function as a bona fide receptor is not yet known. Recent studies inhibiting endogenous GPR30 by expressing antisense cDNA in MCF-7 breast cancer cells indicated that GPR30 had no effect on oestrogen-induced cell proliferation [12], suggesting that GPR30 may not be directly involved in oestrogen signalling. In addition to GPR30, sex-hormone-binding globulin (SHBG), the plasma carrier of sex steroid, has been implicated in rapid extranuclear oestrogen and androgen signalling. A cell-membrane high-affinity receptor (R_{SHBG}) for SHBG has been identified that is reported to transduce a G-protein-dependent steroid activation of adenylate cyclase and the generation of cAMP [24]. However, the mechanism of how various steroid agonists or antagoinsts bound to SHBG elicit different cell signalling responses is not known. Interestingly, the oestradiol-induced proliferation of MCF-7 breast cancer cells was found to be inhibited by SHBG through R_{SHBG} [25].

Oestrogen has been shown to rapidly increase cAMP production, intracellular calcium and intracellular InsP_3 [20,26]. These rapid effects of oestrogen are consistent with G-protein activation through a GPCR. As an alternative model to the existence of a separate or membrane GPCR for oestrogen, studies have suggested that a subpopulation of membrane-localized classical ER may physically associate with and activate G-proteins. For example, ecotopically expressed ERα and ERβ in Chinese hamster ovary cells were shown to physically associate with and activate various G-protein α subunits, including Gα_s and Gα_q [27]. In endothelial cells, endogenous ERα localized to the membrane and associated with Gα_i can mediate activation of eNOS production that is sensitive to pertussis toxin [28]. More recently, Razandi et al. [29] showed that ER localized in the cell membrane, bound to oestrogen and rapidly activated matrix metalloproteinases 2 and 9 through Gα_q, Gα_i, and G$\beta\gamma$-dependent mechanisms. These matrix metalloproteinases are necessary for oestrogen-induced heparin-binding epidermal growth factor (EGF) cleavage, making EGF available to transactivate EGF receptor and activate downstream signalling cascades of EGF receptors, including ERK and PI 3-kinases. These data further suggest the existence of crosstalk between ER and EGF receptor and a biological role for ER in the plasma membrane.

Cell-membrane localization of a subpopulation of classical ERs

There is growing evidence that a subpopulation of classical ER associates with the cell membrane. Ecotopic expression of classical nuclear ER in Chinese hamster ovary cells demonstrated that a small portion of total ER synthesized from the same transcript was localized to the cell membrane [27]. By confocal microscopy, Song et al. [16] also showed that a portion of endogenous ERα in MCF-7 breast cancer cells translocates to areas of membrane ruffles after short-term (15 min) oestrogen treatment. Biochemical isolation of purified plasma membrane from endothelial cells showed the presence of ER in caveolae in response to oestrogen [30]. ER targeted to nuclei failed to mediate oestrogen anti-apoptotic effects in osteoblasts [18] and oestrogen activation of ERK1/2 in Chinese hamster ovary cells [31], while ER targeted to the cell membrane mediated rapid effects of oestrogen similar to that of the wild-type receptor. The ER LBD appears to be the minimum domain required for mediating rapid oestrogen signalling responses. ER LBD targeted to the cell membrane retains its ability to mediate rapid oestrogen activation of ERK [31] and to rescue bone cells from apoptosis similar to full-length ERα [18]. As further evidence that cell membrane/cytoplasmic localization of ER could play a significant biological role, Kumar et al. [32] recently reported that a naturally occurring variant of the metastatic tumour antigen 1 sequestered ER in the cytoplasm of breast cancer cells. The result of this cytoplasmic retention was reduced oestrogen-mediated transcription and enhancement of extranuclear oestrogen-initiated activation of ERK. Interestingly, the expression of metastatic tumour antigen 1 is higher in human breast tumours than in the adjacent normal tissue and a high level of metastatic tumour antigen 1 expression in tumour cells correlated with cytoplasmic localization of ER. These data further indicate that rapid oestrogen-initiated signalling is mediated through extranuclear actions of the receptor and may be of pathophysiological consequence.

How ER localizes to the cell membrane is not well understood. There is no evidence of post-translational modifications that could facilitate membrane insertion of ER, including palmitoylation, myristoylation and glycosylphosphoinositol [33]. Recent evidence suggests that ER translocation to the cell membrane is influenced by caveolin-1 [31]. ER was found to be associated with caveolin-1 in purified plasma membranes and caveolin-1 overexpression facilitated ER translocation to the cell membrane. Razandi et al. [33] recently identified Ser-522 in the LBD of mouse ERα as an important site for efficient membrane localization. A Ser-522→Ala mutation of ERα was significantly less effective in binding to caveolin-1 and localization to the plasma membrane as compared with the wild-type ERα. Furthermore, the Ser-522→Ala mutant behaved as a dominant-negative version of wild-type ER by sequestering and limiting the numbers of ER at the cell membrane, and reducing oestrogen-mediated activation of ERK and oestrogen stimulation of cell-cycle progression [31].

The role of extranuclear ER signalling *in vivo*

Most studies to date on extranuclear oestrogen-initiated signalling have been done in cell-culture systems. Only a few experiments have been reported, suggesting that rapid signalling actions of ER have a role in non-reproductive target tissue *in vivo*. Vascular protection by oestrogen in ischaemia/reperfusion injury *in vivo* requires oestrogen-induced activation of eNOS, as mediated by the PI 3-kinase/Akt pathway [19]. Recently described ER ligands that dissociate the transcription from non-transcription actions of ER have been also identified. A synthetic compound termed oestren (4-oestren-3α,17β-diol) was reported to induce only non-transcriptional activities of ER, whereas another pyrazole compound induces transcriptional activities of ER with minimal effects on rapid signalling activities of ER [34]. The oestren compound administered to mice was found to be as effective as oestradiol in preventing ovariectomy-induced apoptosis of osteoblasts and in preserving bone density, but oestren treatment could not compensate for reduced uterine mass [34]. Additional evidence of an *in vivo* role for rapid oestrogen signalling comes from studies showing that the expression of subsets of endogenous oestrogen-regulated genes is induced through rapid ER activation of Src/ERK and PI 3-kinase/Akt signalling pathways that converge upon and activate other nuclear transcription factors, including c-Fos [35], Elk-1 [36], STAT5 and STAT3 (signal transducer and activator of transcription 5 and 3) [37]. In addition, inhibitors of PI 3-kinase influence a subset of endogenous oestrogen-regulated genes [38]. Thus extranuclear ER signalling could conceivably regulate specific gene networks that either complement or broaden gene repertoires regulated by the direct action of ER in the nucleus as a transcription factor.

Conclusions

Although significant advances in the understanding of the rapid extranuclear signalling actions of oestrogen and progesterone have been made in the past few years, several important questions remain unanswered. Classical intracellular PR can mediate both genomic and extranuclear signalling of progesterone. It will be important in future studies to determine the contribution of extranuclear signalling to biological responses to progesterone such as cell proliferation, adhesion, mobility and migration, and gene expression. The discovery of a novel mPR opens the way for study of the interrelationship between classical PR and mPR. That both mPR and classical PR can mediate rapid progesterone activation of MAPK suggests the possibility of crosstalk between the two receptors. Whether the classical PR and mPR exist in the same tissues or cells and activate the same or distinct signalling pathways remains to be determined. In addition, more work will be needed to determine the biological roles of mPR in higher vertebrates and the role of the different mPR isoforms in different tissues.

Although extranuclear signalling actions of oestrogen have been shown to be mediated in part by classical ER localized to cell membrane, how a sub-population of ER localizes to the cell membrane remains to be determined. Unlike mPR, which has been cloned and characterized, a novel membrane ER has yet to be identified. The identification of MNAR has created exciting opportunities for further study and provides important insights into how ER can couple with and trigger activation of signal transduction pathways. However, further experiments are needed to prove the proposed mechanism of how MNAR physically and functionally couples ER with Src (Figure 2B). How MNAR expression affects ER-mediated cell responses such as cell proliferation, survival and apoptosis also remains to be determined. It will also be important to determine how MNAR expression is regulated and if MNAR expression in breast tumours plays a role in the cellular sensitivity of ER to oestrogen and synthetic ligands. It will be interesting to determine the role of MNAR in mediating rapid signalling actions of other steroid hormones. PR is the only steroid hormone receptor that contains a PXXPXR motif, capable of direct interaction with SH3 domains [6], suggesting that it may not require an adapter such as MNAR and may have evolved a more direct mechanism for interaction with and activation of Src. Although AR has been reported to interact with SH3 of Src through a polyproline sequence in its N-terminal domain, interaction with Src does not appear to be direct. Purified recombinant AR is not able to directly bind SH3 domain of Src [6]. AR has been reported to interact and activate Src indirectly through ER [39].

It will also be important to determine whether MNAR and p160 co-activators compete for interaction of AF-2 through LXXLL motifs and, if so, how this might affect the relative balance between transcriptional and non-transcriptional activities of ER. Could MNAR also play a role in facilitating translocation of a subpopulation of ER to the cell membrane? The intracellular localization of MNAR and how it traffics in the cell with ER also needs to be determined. As we learn more about the mechanism and *in vivo* roles of rapid ER extranuclear-initiated signalling in different tissues, it may be possible to develop ER ligands that selectively target the transcription or non-transcription signalling pathways for different therapeutic purposes. It may turn out that MNAR participates in selectively inducing the non-transcriptional activities of ER through distinct conformational changes in the AF-2 region that favour interaction with MNAR over direct transcriptional co-activators. Finally, since PR is co-expressed with ER in target cells, it will be important to determine whether and how crosstalk between extranuclear PR and ER signalling pathways contribute to the well-known physiological interplay between oestrogen and progesterone in reproductive tissues.

Summary

- *The classical nuclear PR contains a PXXPXR motif in the N-terminal domain that interacts directly with the SH3 domain of Src and activates Src through an SH3-domain-displacement mechanism.*

- *A novel mPR unrelated to nuclear PR has been cloned and characterized from sea trout ovaries. mPR binds progestin with high affinity and is required for progesterone-induced oocyte maturation. mPR appears to be a GPCR with seven transmembrane domains that mediates rapid progesterone-induced inhibition of adenylate cyclase and activation of MAPK.*
- *The classical nuclear ER through an adapter protein, MNAR, interacts with SH3 domain of Src and activates its kinase activity.*
- *A subpopulation of classical ER associated with the cell membrane can mediate certain extranuclear signalling actions of oestrogen. Sequences in the ER LBD appear to be important for membrane localization.*

References

1. Losel, R. & Wehling, M. (2003) Non-genomic actions of steroid hormomes. *Nat. Rev. Mol. Cell. Biol.* **4**, 46–52
2. Vegato, E., Shahbaz, M.M., Wen, D.X., Goldman, M.E., O'Malley, B.W. & McDonnell, D.P. (1993) Human progesterone receptor A form is a cell- and promoter- specific repressor of human progesterone receptor B function. *Mol. Endocrinol.* **7**, 1244–1255
3. Maller, J.L. (2001) The elusive progesterone receptor in *Xenopus* oocytes. *Proc. Natl. Acad. Sci. U.S.A.* **98**, 8–10
4. Tian, J., Kim, S., Heilig, E. & Ruderman, J.V. (2000) Identification of XPR-1, a progesterone receptor required for *Xenopus* oocyte activation. *Proc. Natl. Acad. Sci. U.S.A.* **97**, 14358–14363
5. Bayaa, M., Booth, R.A., Sheng, Y. & Liu, X.J. (2000) The classical progesterone receptor mediates *Xenopus* oocyte maturation through a nongenomic mechansim. *Proc. Natl. Acad. Sci. U.S.A.* **97**, 12607–12612
6. Boonyaratnakornkit, V., Scott, M.P., Ribon, V., Sherman, L., Anderson, S.M., Maller, J.L., Miller, W.T. & Edwards, D.P. (2001) Progesterone receptor contains a proline-rich motif that directly interacts with SH3 domains and activates c-Src familiy tyrosine kinase. *Mol. Cell* **8**, 269–280
7. Bagowski, C.P., Myers, J.W. & Ferrell, Jr, J.E. (2001) The classical progesterone receptor associates with p42 MAPK and is involved in phosphatidylinositol 3-kinase signaling in *Xenopus* oocytes. *J. Biol. Chem.* **276**, 37708–37714
8. Migiliaccio, A., Piccolo, D., Castoria, G., Demenico, M.D., Bilancio, A., Lombardi, M., Gong, W., Beato, M. & Auricchio, F. (1998) Activation of Src/p21ras/Erk pathway by progesterone via cross-talk with estrogen receptor. *EMBO J.* **17**, 2008–2018
9. Xu, W., Doshi, A., Lei, M., Eck, M.J. & Harrison, S.C. (1999) Crystal structures of c-Src reveal features of its autoinhibitory mechanism. *Mol. Cell* **3**, 629–638
10. Zhu, Y., Rice, C.D., Pang, Y., Pace, M. & Thomas, P. (2003) Cloning, expression and characterization of a novel membrane progestin receptor and evidence it is an intermediary in meiotic maturation of fish oocytes. *Proc. Natl. Acad. Sci. U.S.A.* **100**, 2231–2236
11. Zhu, Y., Bond, J. & Thomas, P. (2003) Identification, classification, and partial characterization of genes in humans and other vertebrates homologous to a fish membrane progestin receptor. *Proc. Natl. Acad. Sci. U.S.A.* **100**, 2237–2242
12. Ahola, T.M., Manninen, T., Alokio, N. & Ylikomi, T. (2002) G protein-coupled receptor 30 is critical for a progestin-induced growth inhibition in MCF-7 breast cancer cells. *Endocrinology* **143**, 3376–3384
13. Couse, J.F. & Korach, K.S. (1999) Estrogen receptor null mice: what have we learned and where will they lead us? *Endocr. Rev.* **20**, 358–417

14. Migliaccio, A., Domenico, M.D., Castoria, G., Falco, A., Bontempo, P., Nula, E. & Auricchio, F. (1996) Tyrosine kinase/p21ras/MAP-kinase pathway activation by estradiol-receptor complex in MCF-7 cells. *EMBO J.* **15**, 1292–1300
15. Castoria, G., Barone, M.V., Demenico, M.D., Bilancio, A., Ametrano, D., Migliaccio, A. & Auricchio, F. (1999) Non-transcriptional action of estradiol and progestin triggers DNA synthesis. *EMBO J.* **18**, 2500–2510
16. Song, R.X.-D., McPherson, R.A., Adam, L., Bao, Y., Shupnik, M., Kumar, R. & Santen, R.J. (2002) Linkage of rapid estrogen action to MAPK activation of ERα-Shc association and Shc pathway activation. *Mol. Endocrinol.* **16**, 116–127
17. Castoria, G., Migliaccio, A., Domenico, M.D., Bilancio, A., Falco, A.D., Lombardi, R.,. Fiorentino, R., Varrichio, M., Barone, M.V. & Auricchio, F. (2001) PI3 kinase in concert with Src promotes the S-phase entry of oestradiol-stimulated MCF-7 cells. *EMBO J.* **20**, 6050–6059
18. Kousteni, S., Bellido, T., Plotkin, L.I., O'Brien, C.A., Bodenner, D.L., Han, L., Han, K., Digregorio, G.B., Katzenellenbogen, J.A., Katzenellenbogen, B.S., Roberson, P.K., Weinstein, R.S., Jilka, R.L. & Manolagas, S.C. (2001) Nongenotropic, sex-nonspecific signaling through the estrogen or androgen recptors: dissociation from transcriptional activity. *Cell* **104**, 719–730
19. Simoncini, T., Hafezi-Moghadam, A., Brazil, D.P., Ley, K., Chin, W.W. & Liao, J.K. (2000) Interaction of oestrogen receptor with the regulatory subunit of phosphatidylinositol-3-OH kinase. *Nature (London)* **407**, 538–541
20. Levin, E.R. (2002) Cellular functions of plasma membrane estrogen receptors. *Steroids* **67**, 471–475
21. Wong, C.-W., McNally, C., Nickbarg, E., Komm, B.S. & Cheskis, B.J. (2002) Estrogen receptor-interacting protein that modulates its nongenomic activity-crosstalk with Src/Erk phosphorylation cascade. *Proc. Natl. Acad. Sci. U.S.A.* **99**, 14783–14788
22. McKenna, N.J. & O'Malley, B.W. (2002) Combinational control of gene expression by nuclear receptors and coregulators. *Cell* **108**, 465–474
23. Bartletta, F., Chi-Wai, W., McNally, C., Komm, B., Katzenellenbogen, B.S. & Cheskis, B.J. (2004) Characterization of the interactions of the estrogen receptor and MNAR in activation of cSCR. *Mol. Endocrinol.*, in the press
23a. Filardo, E.J., Quinn, J.A., Bland, K.I. & Frackelton, A.R.J. (2000) Estrogen-induced activation of Erk-1 and Erk-2 requires the G protein-coupled receptor homolog, GPR30, and occurs via *trans*-activation of the epidermal growth factor receptor through release of HB-EGF. *Mol. Endocrinol.* **14**, 1649–1660
24. Rosner, W., Hryb, D.J., Khan, M.S., Nakhla, A.M. & Romas, N.A. (1999) Sex hormone-binding globulin mediates steroid hormone signal transduction at the plasma membrane. *J. Steroid Biochem. Mol. Biol.* **69**, 481–485
25. Fortunati, N., Becchis, M., Catalano, M., Comba, A., Ferrera, P., Raineri, M., Berta, L. & Frairia, R. (1999) Sex hormone-binding globulin, its membrane receptor, and breast cancer: a new approach to the modulation of estradiol action in neoplastic cells. *J. Steroid Biochem. Mol. Biol.* **69**, 473–479
26. Watson, C.S. & Gametchu, B. (1999) Membrane-initiated steroid actions and the proteins that mediate them. *Proc. Soc. Exp. Biol. Med.* **220**, 9–19
27. Razandi, M., Pedram, A., Greene, G.L. & Levin, E.R. (1999) Cell membrane and nuclear estrogen receptors drive from a single transcript: studies of ERα and ERβ expressed in CHO cells. *Mol. Endocrinol.* **13**, 307–319
28. Wyckoff, M.H., Chambliss, K.L., Mineo, C., Yuhama, I.S., Mendelsohn, M.E., Mumby, S.M. & Shaul, P.W. (2001) Plasma membrane estrogen receptors are coupled to endothlial nitric-oxide synthase through Gαi. *J. Biol. Chem.* **276**, 27071–27076
29. Razandi, M., Pedram, A., Park, S.T. & Levin, E.R. (2003) Proximal events in signaling by plasma membrane estrogen receptors. *J. Biol. Chem.* **278**, 2701–2712
30. Kim, H.P., Lee, J.Y., Leong, J.K., Bae, S.W., Lee, H.K. & Jo, I. (1999) Nongenomic stimulation of nitric oxide release by estrogen is mediated by estrogen receptor α localized in caveolae. *Biochem. Biophys. Res. Commun.* **263**, 257–262

31. Razandi, M., Oh, P., Pedram, A., Schnitzer, J. & Levin, E.R. (2002) ERs associates with and regulate the production of caveolin: implications for signaling and cellular actions. *Mol. Endocrinol.* **16**, 100–115
32. Kumar, R., Wang, R.-A., Mazumdar, A., Talukder, A.H., Mandai, M., Yang, Z., Bagheri-Yarmand, R., Sahin, A., Hortobagyi, G., Adam, L. et al. (2002) A naturally occurring MTA1 variant sequesters oestrogen receptor-α in the cytoplasm. *Nature (London)* **418**, 654–657
33. Razandi, M., Alton, G., Pedram, A., Ghonshani, S., Webb, P. & Levin, E.R. (2003) Identification of structural determinant necessary for the localization and fuction of estrogen receptor α at the plasma membrane. *Mol. Cell. Biol.* **23**, 1633–1646
34. Kousteni, S., Chen, J.R., Bellido, T., Han, L., Ali, A.A., O'Brien, C.A. Plotkin, L.I., Fu, Q., Mancino, A.T., Weinstein, R.S. et al. (2002) Reversal of bone loss in mice by non-genotropic signaling of sex steroids. *Science* **298**, 843–846
35. Daun, R., Xie, W., Li, X., McDougal, A. & Safe, S. (2002) Estrogen regulation of c-fos gene expression through phosphotidylinositol-3-kinase dependent activation of serum response factor in MCF-7 breast cancer cells. *Biochem. Biophys. Res. Commun.* **294**, 384–394
36. Daun, R., Xie, W., Burghardt, R.C. & Safe, S. (2001) Estrogen receptor-mediated activation of serum response element in MCF-7 cells through MAPK-dependent phosphorylation of Elk-1. *J. Biol. Chem.* **276**, 11590–11598
37. Bjornstrom, L. & Sjoberg, M. (2002) Signal transducers and activator of transcription as downstream targets of nongenomic estrogen receptor actions. *Mol. Endocrinol.* **16**, 2202–2214
38. Pedram, A., Razandi, M., Aitkenhead, M., Hughes, C.W. & Levin, E.R. (2002) Integration of the non-genomic and genomic actions of estrogen. Membrane-initiated signaling by steroid to transcription and cell biology. *J. Biol. Chem.* **277**, 50768–50775
39. Migliaccio, A., Castoria, G., Di Domenico, M., De Falco, A., Bilancio, A., Lombardi, M., Barone, M.V., Ametrano, D., Zannini, M.S., Abbondanza, C. & Auricchio, F. (2000) Steroid-induced androgen receptor-oestradiol receptor-β-Src complex triggers prostate cancer cell proliferation. *EMBO J.* **19**, 5406–5417

9

Nuclear receptors and disease: androgen receptor

Bruce Gottlieb*†[1], Lenore K. Beitel*‡, Jianhui Wu*, Youssef A. Elhaji*§ and Mark Trifiro*‡§

**Lady Davis Institute for Medical Research, Sir Mortimer B. Davis-Jewish General Hospital, 3755 Chemin de la Côte-Ste-Catherine, Montréal, Québec, Canada H3T 2E1, †Department of Biology, John Abbott College, P.O. Box 2000, Ste Anne de Bellevue, Québec, Canada H9X 3L9, ‡Department of Medicine, McGill University, Montréal, Québec, Canada H3T 1E2, and §Department of Human Genetics, McGill University, Montréal, Québec, Canada H3T 1E2*

Abstract

The androgen receptor (AR) protein regulates transcription of certain genes. Usually this depends upon a central DNA-binding domain that permits the binding of androgen–AR complexes to regulatory DNA sequences near or in a target gene. The AR also has a C-terminal ligand-binding domain and an N-terminal transcription modulatory domain. These N- and C-terminal domains interact directly, and with co-regulatory, non-receptor proteins, to exert precise control over a gene's transcription rate. The precise roles of these proteins are active research areas. Severe X-linked AR gene (*AR*) mutations cause complete androgen insensitivity, mild ones impair virilization with or without infertility, and moderate ones yield a wide phenotypic spectrum sometimes among siblings. Different phenotype expressivity may reflect variability of AR-interactive proteins. Mutations occur throughout the *AR* but are concentrated in specific areas of the gene known as hot spots. A number of these mutations

[1]*To whom correspondence should be addressed, at the Lady Davis Institute for Medical Research (e-mail bruce.gottlieb@mcgill.ca).*

of somatic origin are associated with prostate cancer. N-terminal polyglutamine (polyGln) tract expansion reduces AR transactivation, and when there are more than 38 glutamine residues it causes spinobulbar muscular atrophy, a motor neuron disease, due to a gain of function. Variations in polyGln tract length have been associated as risk factors with prostate, breast, uterine, endometrial and colorectal cancer, as well as male infertility.

Introduction

The androgen receptor [AR; Online Mendelian Inheritance in Man (OMIM) code 31370] is a member of the superfamily of nuclear receptors that function as ligand-dependent transcription factors. The gene for the AR is approx. 90 kb, with eight exons, and lies on the X chromosome at Xq11–12. Like other nuclear receptors, the AR protein contains four major domains, the N-terminal domain, the DNA-binding domain (DBD), the hinge region and the androgen- or ligand-binding domain (LBD). The DBD and LBD show considerable homology with other nuclear receptors. The DBD contains two zinc fingers and is required for androgen-response-element recognition. The 253 residue C-terminal LBD contains 12 α-helices and a highly hydrophobic ligand-binding site. Intracellular AR is essential for androgen action, whether the ligand is testosterone or its 5α-reduced derivative, 5α-dihydrotestosterone. Hence, the AR is essential for normal primary male sexual development before birth (masculinization), and for normal secondary male sexual development around puberty (virilization). AR dysfunctions in XY individuals result in androgen-insensitivity syndromes (AISs).

Structure–function relationships of the AR protein and gene

In the cell cytoplasm the unliganded AR is associated with a number of heat-shock proteins (hsps). The AR is activated by the binding of an androgen molecule, usually 5α-dihydrotestosterone, and the release of hsps. Ligand binding also promotes AR hyperphosphorylation. The transformed AR assumes an altered conformation in which two zinc fingers in the DBD are exposed (Figure 1). It is then translocated into the nucleus, where homodimerization occurs and the AR homodimer binds to specific androgen-response elements in certain genes acquiring the ability to regulate the rate of transcription of these genes (Figure 1). To exert such regulation, a complex of an androgen and an AR must also interact with transcriptionally active proteins that bind upstream and downstream of the androgen-response

Figure 1. Mechanism of androgen action in the cell from ligand to gene activation to protein expression
T, testosterone; DHT, 5α-dihydrotestosterone; N-term, N-terminal region of AR protein; DBD, DBD of AR gene, LBD, LBD of AR gene; hsps, heat-shock proteins associated with AR gene; P, phosphorylation site; RNA Pol II, RNA polymerase II. These mechanisms are described in a number of papers [1,2,7,10].

☞

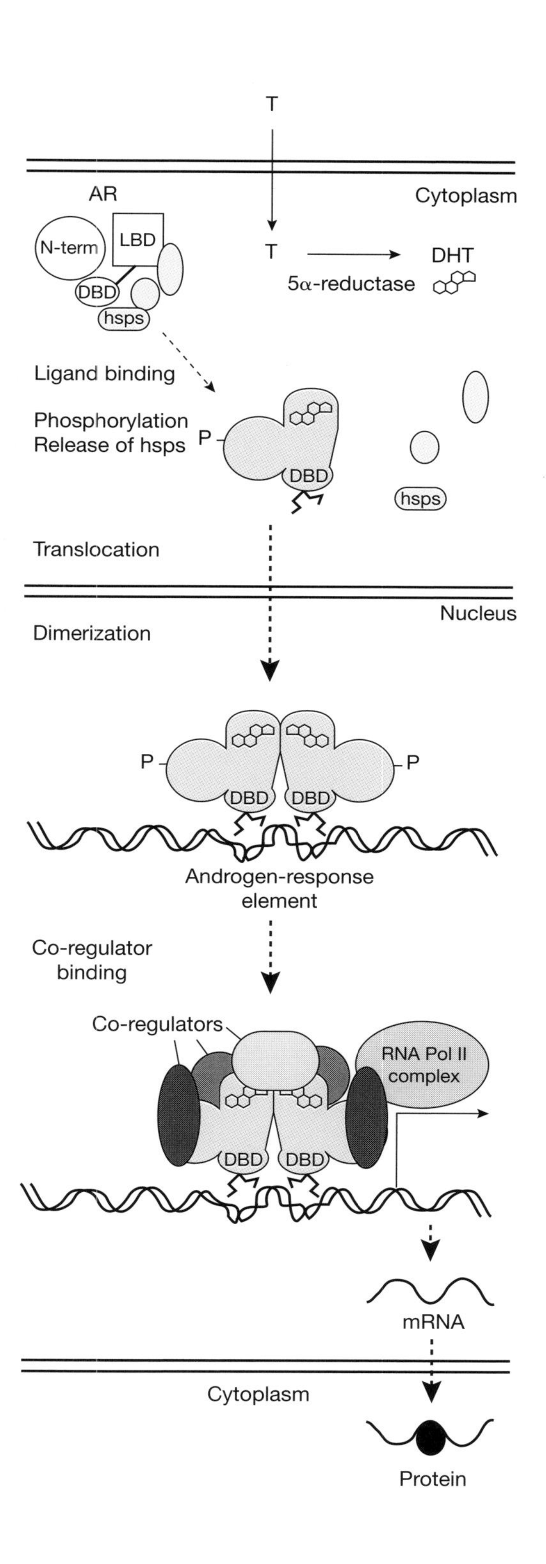
T
AR
Cytoplasm
N-term
LBD
T
DHT
DBD
5α-reductase
hsps
Ligand binding
Phosphorylation
Release of hsps
P
DBD
hsps
Translocation
Nucleus
Dimerization
P
P
DBD
DBD
Androgen-response
element
Co-regulator
binding
Co-regulators
RNA Pol II
complex
DBD
DBD
mRNA
Cytoplasm
Protein

element. These interactions, with a number of co-regulators, including basal transcription factors and core promoter element-binding proteins, determine the vectorial control over the transcriptional expression of a given androgen target gene. The mechanism of action of the AR are reviewed in a recent paper by Gobinet et al. [1]. The nature and possible function of these co-regulators have been reviewed by Heinlein and Chang [2], and a comprehensive database of these co-regulatory proteins is available in the AR Gene Mutations Database (http://www.mcgill.ca/androgendb).

While the *AR* is about 90 kb long, only approx. 2.75 kb, divided into eight exons, codes for amino acids. The arrangement of the exons and introns is shown in Figure 2. The variable length of the AR protein reflects the fact that its N-terminal transregulation modulatory portion (approx. 537 amino acids) contains two homopolymeric amino acid 'repeats' that are polymorphic in size (Figure 2): the polyglutamine (polyGln) repeat size varies from 11 to 37 [3];

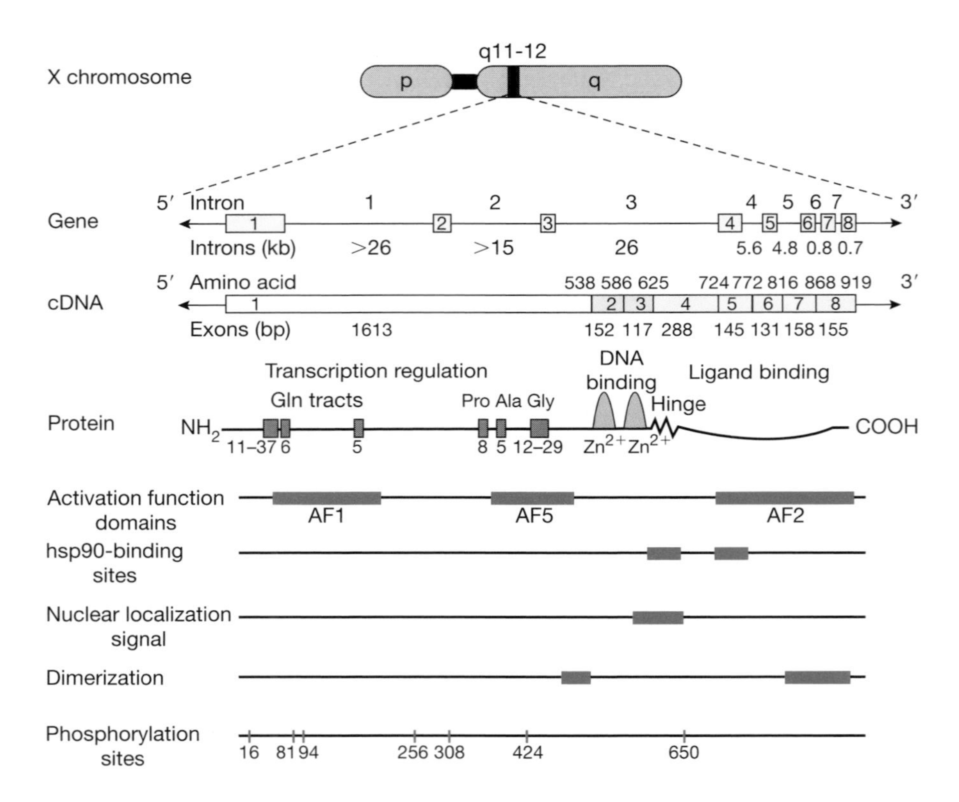

Figure 2. Major structure–function domains and putative subdomains of the AR gene and protein

Note that the thicker portions of the lines below the protein structure indicate the most likely location of a given functional domain, with the exception of the phosphorylation sites, which are at specific numbered residues. Zn^{2+} refers to the zinc fingers; AF refers to the transcriptional activation function regions. The nature of these domains is reviewed in a number of papers [1,7,10].

the other, polyglycine, varies from 12 to 29 [4]. The central DBD (approx. amino acids 557–616) is encoded by exons 2 and 3. Adjacent to the DBD, C-terminally, there is a hinge region, (amino acids 617–663) encoded by exons 3 and 4. Finally, the C-terminal LBD (approx. 253 amino acids, starting at amino acid 664) is encoded by exons 4–8. In addition to their principal functions, the LBD, DBD, hinge and N-terminal domains embody subsidiary functions that affect dimerization (involving N- and C-terminal and LBD–LBD interactions), nuclear localization, transcriptional regulation, hsp binding and phosphorylation [1] (Figure 2). In addition, a number of activation function domains (AF) has been identified (Figure 2). Thus the tetramodular (domain) concept of AR function is a simplification; instead, domain interaction, and interaction with co-regulatory proteins, hold the secrets to a full understanding of an AR's structure–function properties.

Diseases as a result of mutations in the AR gene

Loss-of-function diseases: AISs

AISs (OMIM # 300068) can be subdivided into three phenotypes: complete AIS (CAIS), partial AIS (PAIS) and mild AIS (MAIS). *AR* mutations that severely impair the amount, structure or function of the AR cause the complete androgen insensitivity phenotype. Standard references quote rates of 2–5 in 100 000 for complete androgen insensitivity. Subjects are born looking unambiguously female because 5α-dihydrotestosterone-dependent masculinization of the external genital primordia is totally absent. These individuals are typically not suspected of being abnormal until the onset of puberty, when breast development is normal, but pubic and axillary hair are not. Menarche, initially considered 'late', never occurs. Müllerian duct regression, being androgen-independent, is normal. Their testes may or may not be inguinal.

Because all XY subjects with complete androgen insensitivity are sterile (genetic lethals), one-third of their mutant alleles should represent new mutations. A recent report on single-case families with complete or partial androgen insensitivity gene mutations [5] revealed a *de novo AR* mutation rate of close to 30% (8 out of 30), thereby affirming the theoretical expectation of 33% for an X-linked recessive genetic lethal [6].

Partial androgen insensitivity has a highly variable phenotypic expression. At one extreme, the external genitalia are near-normal female, except for clitoromegaly and/or posterior labial fusion; at the other extreme, the genitalia may be morphologically normal male, but small, or there may be simple coronal hypospadias or a prominent midline raphe of the scrotum. In between these extremes are all grades of frank external genital ambiguity that are, nonetheless, predominantly masculine or predominantly feminine. Indeed, for some mutations in the LBD, such variable expressivity may be the rule, not the exception

[7]. Furthermore, in rare families with partial androgen insensitivity, the expressivity may vary markedly, from near-normal male to near-normal female.

Mild androgen insensitivity takes two phenotypic forms at puberty: in one, spermatogenesis and fertility are impaired [8]; in the other, spermatogenesis is normal, or sufficient to preserve fertility [9]. In both, gynaecomastia, high-pitched voice, sparse sex hair and impotence may be noted. In the form where fertility is preserved, one presumes that the dysfunction of the mutant AR is sufficiently mild so that it can be overcome by collaboration with the set of co-regulatory proteins that is active in Sertoli cells [10].

Phenotype–genotype correlation of *AR* mutations

The present version of the Androgen Receptor Gene Mutation Database [11] (see http://www.mcgill.ca/androgendb), contains 450 entries of mutations causing AIS representing over 300 different *AR* mutations from more than 600 patients with AIS and demonstrates an unequal distribution of these mutations along the length of the *AR*, as shown in Table 1. It has previously been suggested that mutation-dense regions are hot spots that reflect the high density of mutable CpG sites in the region [12]. It is also apparent that the types of mutations differ along the length of the *AR*. In particular, nearly all mutations in exon 1 (Table 1) cause complete androgen insensitivity, and nearly all are of the premature translation termination variety, whether by direct mutation to a stop codon or indirectly by frameshifts after small deletions or insertions. To date, only 54 mutations have been reported in exon 1 of the AR in patients suffering from some form of AIS, despite the fact that it encodes more than half of the AR protein [11], and even fewer in splicing and untranslated regions of the AR gene (Table 1).

In the LBD there is a striking preponderance of missense mutations with a significantly greater number of CAIS than PAIS cases (Table 1). In an effort to better understand the structure–function relationship of how specific mutations in the LBD cause AIS, the X-ray crystal structure of the AR LBD was solved [13]. This revealed a structure that consisted of 12 α-helices (Figure 3). However, the putative crystal structure only identified 18 specific residues in the LBD that were in close contact with the ligand-binding pocket when modelled with the synthetic ligand R1881. This meant that the vast majority of AIS mutants whose ARs exhibited reduced or even no ligand binding could not be explained by the crystal structure. A selection of CAIS, PAIS and MAIS mutations are shown in Figure 3, which shows that the nature of their AIS phenotypes has little to do with their distance from the ligand-binding pocket. Therefore, to try to elucidate how such mutations could effect the structure of the ligand-binding pocket, and so explain the lack of ligand binding in these mutants, we have recently used molecular dynamic modelling techniques over extended periods of time (up to 4 ns) to, in effect, create four-dimensional structures of AR mutants [14].

Table 1. Nature and distribution of unique AR mutations that cause disease

N-terminal domain, amino acids 1–534; DBD, amino acids 559–624; Hinge region, amino acids between DBD and LBD 625–663; LBD, amino acids 664–919. UTR, untranslated region.

Loss-of-function disease	Type of mutation	Number of mutations					
		N-terminal domain	DBD	Hinge region	LBD	Splice site	Intron
CAIS	Single-base substitution	4	15		87	12	1
	Premature termination	23	2	1	13		
	Complete gene deletion	3					
	Partial gene deletion	6	4		3		
	Deletion (1–4 bases)	8	3				
	Insertion	5				1	
	Duplication	2					
PAIS	Single-base substitution	3	18	1	60	2	
	Multiple-base substitution		1		1		
	Premature termination	1					
	Deletion (1–4 bases)				1		1
MAIS	Single-base substitution	7			10		
	Partial gene deletion				1 (?)		
	Deletion (1–4 bases)				1		

continued ☞

Table 1. (contd)

Loss-of-function disease	Type of mutation	Number of mutations					
		N-terminal domain	DBD	Hinge region	LBD	Splice site	UTR
Prostate cancer	Single-base substitution	25	6	2	38		2
	Premature termination	1			2		
	Deletion (1–4 bases)	5			1		
	Insertion					1	1
Breast cancer	Single base substitution		2			1	
Larynx cancer	Deletion (30 bases)	1					

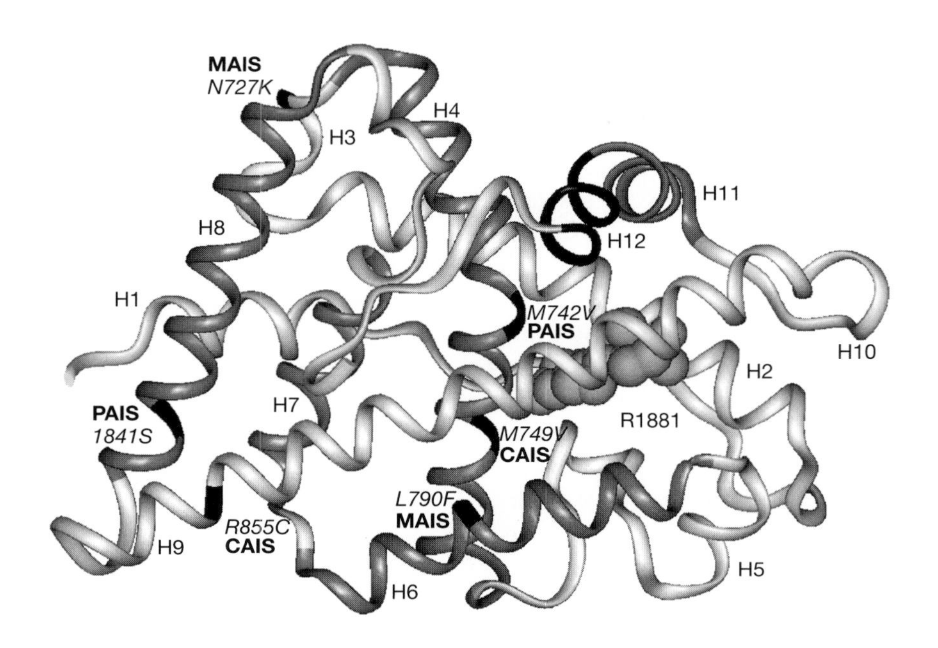

Figure 3. The AR LBD as defined by the X-ray crystal structure (Protein Data Bank code 1e3g)
The α-helices (H) are numbered and spheres illustrate the bound ligand, R1881. Six specific mutations (in italic) together with their AIS phenotype (in bold) are shown to illustrate that mutations affecting ligand binding can occur both near to and far away from the ligand-binding pocket. Distances between c-α and the bound ligand are as follows. CAIS, Arg-855, 26.3 Å; Met-749, 4.1 Å. PAIS, Met-742, 7.0 Å; Ile-841, 25.1 Å. MAIS, Leu-790, 12.5 Å; Asn-727, 27.3 Å. 1Å = 0.1 nm

In these studies a CAIS mutant not close to the ligand-binding pocket causes a local structural distortion that also affects the ligand-binding pocket. Thus, modelling has resulted in a better understanding of how alterations in residues that are not close to the putative binding pocket affect ligand binding. In future, such studies could ultimately be used to create androgen analogues with which to treat AIS patients.

The issue of variable expressivity is one of the most puzzling in trying to understand the genotype–phenotype relationship of many LBD mutations. The traditional explanation for such variable expressivity of an AR mutation is that the level or competence of co-regulatory proteins act as genetic 'background' factors in determining the overall clinical outcome [7]. Recently, however, it has been appreciated that somatic mosaicism (more or less covert), may account for some variable expressivity [15]. The simplest origin of such mosaicism would be forward mutation of an inherited normal allele to a mutant allele in a subject with a negative family history. However, in a family with multiple affected individuals, a relatively mild clinical outcome could reflect a back mutation of an inherited mutation to a normal allele.

Gain-of-function diseases

Prostate cancer (CaP)

To date 85 *AR* mutations have been found in CaP tissue, almost all being single-base substitutions due to somatic rather than germline mutations, and the role of the AR in CaP has recently been reviewed by Debes and Tindall [16]. As can be seen in Table 1, while the majority occur in the LBD (approx. 45%), a substantial number occur in exon 1 (approx. 30%). Originally it was thought that AR was not expressed in CaP, but this does not appear to be the case. In fact, the level of AR expression is significantly higher in hormone-resistant tumours compared with hormone-sensitive tumours [17]. Considerable controversy has revolved around conflicting studies that only sometimes find a significant number of AR mutations in CaP [18]. It has been argued that AR mutations only appear during the latter stages of CaP and, in addition, some studies have indicated that anti-androgen treatments have resulted in AR mutations. There is evidence from study of somatic *AR* mutations in prostatic carcinoma that certain ones of the missense variety not only permit the AR to bind unusual androgens or other steroids [19], promiscuously, but also allow it to be activated by them [20] in a manner that allows these unorthodox steroid–receptor complexes to be effective in transcriptional regulation of certain androgen target genes. Further, recent studies have found that a significant percentage of recurrent CaP following treatment have somatic mutations in their *AR*s and that the localization of mutations are influenced by the type of treatment [21]. However, by far the most significant event from a clinical prospective is that all CaP tumours eventually become androgen-resistant or -independent [22], and that in a number of cases this coincides with the appearance of mutations in the *AR*.

Male breast cancer

To date only two cases of *AR* mutations associated with male breast cancer have been reported and in both cases they have been in individuals suffering from PAIS [23]; the mutations being in adjacent codons in exon 3. It therefore seems likely that the *AR* mutations are not a primary cause of male breast cancer, which rather may possibly be due to the increased incidence of gynaecomastia associated with AIS.

Diseases directly associated with *AR* CAG repeat-length variation: spinobulbar muscular atrophy (SBMA; Kennedy disease)

Kennedy disease (OMIM no. 31320) is a spinobulbar motor neuronopathy associated with mild androgen insensitivity, and is one of the classic trinucleotide-repeat-expansion diseases that cause inherited neurogenerative disorders [24]. It is caused by expansion of the glutamine-coding $(CAG)_{8-35}$ CAA tract in exon 1 of the *AR* to a total of 38 or more [25]. It is an adult-onset motor neuronopathy that typically causes slowly progressive, symmetric wasting and weakness, initially of the proximal muscles of the hip and shoulder. Muscle cramps, hand tremors and fasciculations are often associated. Eventually, motor neurons of the brainstem become involved, leading to speech and swallowing difficulties. Male hypogonadism, usually represented by gynaecomastia and testicular atrophy, and attributable to mild androgen insensitivity, is not infrequent. The mild androgen-insensitivity component of SBMA may reflect a loss of AR transcriptional regulatory activity by virtue of a pathologically expanded polyGln tract. It should be noted that, in SBMA, the androgen-insensitivity phenotype is quite variable.

Since subjects with complete androgen insensitivity, including those with complete *AR* deletions, do not develop SBMA, this knowledge mandated the logic that the polyCAG-expanded *AR* or the polyGln-expanded AR protein is somehow motor neuronotoxic by a gain, not a loss, of function.

It is now clear that the polyGln-expanded AR protein, and probably the expanded polyGln tracts themselves, with certain flanking segments of the various parental proteins, are the essential pathogenetic agents. Expansion to beyond approx. 38 repeats endows a polyGln tract with a threshold property (or more than one) that must be selectively lethal to certain neurons. The biochemical, histopathological and neurophysiological features of SBMA are, unremarkably, those secondary to motor denervation. A number of possible causes for this gain of function are listed in Table 2, and these were analysed and discussed in a recent review by La Spada and Taylor [24].

The mild androgen-insensitivity component of SBMA is clearly due to a loss of function by the polyGln-expanded AR protein. Since androgens are both motor neurono*tropic* and motor neurono*trophic*, it is possible that the polyGln-expanded AR protein loses a function that is necessary, but not sufficient, for the motor neuronopathy of SBMA generally or for death of certain motor neurons specifically.

The polyGln-expanded proteins or fragments thereof may oligomerize, directly or through intermediates, either non-covalently (by hydrogen-bonded polar zippering) or covalently (by transglutaminase-catalysed isodipeptide formation), to yield inclusions (aggregates) that accumulate in or around the nucleus of certain neurons [25]. Why neurons are selectively vulnerable to the toxicity of polyGln-expanded proteins when these proteins (e.g. the AR) are widely distributed in many non-neuronal cells and why only certain motor neurons are affected remains entirely speculative.

AR CAG tract-length variation as a risk factor for disease

Female breast cancer

Previous investigations into the relationship of CAG repeat lengths in the *AR* with female breast cancer have yielded somewhat confusing results [26]. There

Table 2. Diseases associated with AR CAG-tract-length variation

Direct association	CAG tract	Androgen sensitivity	Gain of function: possible causes	Symptoms
SBMA	>38	Reduced	1. Misfolding	1. Adult-onset motor neuropathy of proximal hip and shoulder muscles
			2. Truncation	2. Hypogonadism results in gynaecomastia and testicular atrophy
			3. Aggregation	
			4. Sequestration of AR protein/ transcription factors	
			5. Proteasome inhibition	
			6. Mitochondrial dysfunction	

Indirect association	CAG tract	Androgen sensitivity	Associated risk factors	Comments
CaP	Shorter length	Increased	Ethnicity and family history	Inconclusive studies: possible somatic alterations
Male infertility	Longer length	Reduced	Ethnicity	
Female breast cancer	Longer length	Reduced	BRCA1 mutation carriers	Inconclusive studies: possible somatic alterations
Endometrial cancer	Longer length	Reduced		Somatic alterations
Colon cancer	Shorter length	Increased		Selective growth advantage: somatic alterations

is also a suggestion that CAG repeats might be a significant modifier to the BRCA1 mutation associated with breast cancer risk [27], although recently even this observation has been questioned [28]. Interestingly, decreased AR transactivational activity lowers androgen/oestrogen balance, and may thereby effect functional hyperoestrogenicity. This may promote the pathogenesis of breast cancer. To elucidate whether longer CAG repeats of the *AR*, which correlate with lower transactivational activity of the AR, are associated with breast cancer in women over 40, we have examined the distribution of CAG repeat lengths in breast cancer tissue from this population [29]. The breast cancer tissue was histologically graded as grade 1, or well differentiated, grade 2, or moderately differentiated, and grade 3, or poorly differentiated. Analysis showed significant differences as compared with controls when CAG lengths greater than 21 were examined, and that alleles with 26 repeats or more were 2.4-fold more frequent in breast cancer samples than in constitutional samples from a normal population. A significant shift to greater CAG repeat lengths appeared in grade 1 and 2 tumours. Our results give some indication as to the progression of breast cancer by suggesting that hypotransactive ARs with long polyGln tracts may have a role in the initiation and/or progression of breast cancer, which has also been noted in a recent review [30]. In addition grade 3 tumours, perhaps due to increased genomic instability, tended to have shorter than normal CAG repeat lengths. In this case it is hypothesized that the ARs have now become hypertransactive, possibly coinciding with the oestrogen resistance that is associated with grade 3 tumours. Whether this shift is of germline or somatic origin was not clear, though the appearance in 14% of the breast cancer samples of a third CAG repeat length indicates that it may be somatic [26].

Male infertility

In Singapore, otherwise normal males with 28 or more CAG repeats in their *AR* have been reported to have more than a 4-fold increased risk of impaired spermatogenesis, and the more severe the spermatogenic defect, the greater the chance of finding a long repeat [31]. While some additional studies have supported this observation that longer CAG repeats are associated with infertility [32], not all have done so [33].

CaP

Due to androgen activity being inversely proportional to polyGln tract length [34], there has been considerable speculation as to a possible relationship between tract length and CaP. The only significant association seems to be related to the fact that the length of the tract shows ethnic variation, and that this may be one cause for the higher risk of CaP development in, for example, African-Americans [35]. However, to date, the results have been largely inconclusive, as reported in an extensive review by Ferro et al. [26]. Their overall conclusion is that only in combination with other polymorphisms such as that of the prostate-specific antigen (PSA) protein could CAG-tract-length variation in the AR be considered a significant risk factor. To date, however, the most severe limitations of the studies is that they have not examined somatic alterations in *AR* CAG repeats in CaP tissues, as previous studies have shown that *AR* CAG repeats have a high degree of somatic instability [36]. It is hoped that with the ability to examine repeat length in very specific CaP tissues from specific prostate tumours, a better understanding of any possible relationship will emerge.

Uterine endometrial cancer

Only one study to date has examined the possible effect of a CAG-repeat expansion as a risk factor, which may reduce the antagonist effect of the androgens in counteracting the proliferative effect of oestrogens [37].

Colorectal cancer

Recently it has been shown that in 10% of colon cancer samples there was a somatic *AR* CAG-repeat reduction [26]. The exact mechanism of how such instability could lead to tumour growth was strictly the subject of speculation, although the suggestion was made that it might be due to such cells having a selective growth advantage [26].

Conclusions

The ability to correlate specific gene alterations with both phenotype and structure–function of the AR is likely to lead to considerable insight into how these alterations directly cause the expression of disease phenotypes. This is likely to be accomplished using a number of new techniques such as laser capture microdissection and four-dimensional protein modelling. Of particular interest will be our ability to begin to explain such concepts as phenotypic variable expressivity, gain-of-function disease phenotypes and the influence on phenotype of somatic versus germline gene alterations. Finally, the importance of somatic alterations in polymorphic trinucleotide repeats, as risk factors for certain diseases, is likely to be further elucidated.

Summary

- *Gene alterations in the AR cause AIS with a range of phenotypic expression from complete to mild AIS.*
- *Some individual AR alterations have a significant degree of variable expressivity which in some cases is caused by somatic mosaicism.*
- *Somatic AR alterations have been found in CaP tissues in which the AR protein appears to have an acquired gain of function.*
- *Expansion of a CAG repeat in exon 1 of the AR results in the motor neuron disease SBMA due to a gain of function of the AR.*

- *Variations in the length of the AR CAG tract have been identified as a possible risk factor for prostate, female breast, uterine endometrial and colon cancer, as well as male infertility.*

We thank the Canadian Institutes of Health Research for supporting our own work on *AR* mutations and the various diseases associated with them.

References

1. Gobinet, J., Pujol, N. & Sultan, C. (2002) Molecular action of androgens. *Mol. Cell. Endocrinol.* **198**, 15–24
2. Heinlein, C.A. & Chang, C. (2002) Androgen receptor (AR) coregulators: an overview. *Endocr. Rev.* **23**, 175–200
3. Andrew, S.E., Goldberg, Y.P. & Hayden, M.R. (1997) Rethinking genotype and phenotype correlations in polyglutamine expansion disorders. *Hum. Mol. Genet.* **6**, 2005–2010
4. Lumbroso, R., Beitel, L.K., Vasiliou, D.M., Trifiro, M.A. & Pinsky, L. (1997) Codon-usage variants in the polymorphic $(GGN)_n$ trinucleotide repeat of the human androgen receptor gene. *Hum. Genet.* **101**, 43–46
5. Hiort, O., Sinnecker, G.H.G., Holterhus, P.M., Nitsche, E.M. & Kruse, K. (1998) Inherited and *de novo* androgen receptor gene mutations: investigation of single-case families. *J. Pediatr.* **132**, 939–943
6. Haldane, J.B.S. (1935) The rate of spontaneous mutation of a human gene. *J. Genet.* **31**. 317–326
7. Pinsky, L., Beitel, L.K., Kazemi-Esfarjani, P., Lumbroso, R., Vasiliou, D.M., Shkolny, D., Abdullah, A.A.R., Gottlieb, B. & Trifiro, M.A. (1996) Lessons from androgen receptor gene mutations that cause androgen resistance in humans. In Sex Differentiation: Clinical and Biological Aspects, Frontiers in Endocrinology, vol. 20 (Hughes, I.A., ed.), pp. 95–114, Serono Symposia Series, Rome
8. Migeon, C.J., Brown, T.R., Lanes, R., Palacios, A., Amrhein, J.A. & Schoen, E.J. (1984) A clinical syndrome of mild androgen insensitivity. *J. Clin. Endocrinol. Metab.* **59**, 672–678
9. Pinsky, L., Kaufman, K. & Killinger, D.W. (1989) Impaired spermatogenesis is not an obligate expression of receptor-defective androgen resistance. *Am. J. Med. Genet.* **32**, 100–104
10. Gottlieb, B., Pinsky, L., Beitel, L.K. & Trifiro, M. (1999) Androgen insensitivity. *Am. J. Med. Genet.* **89**, 210–217
11. Gottlieb, B., Beitel, L.K., Lumbroso, R., Pinsky, L. & Trifiro, M. (1999) Update of the androgen receptor gene mutations database. *Hum. Mutat.* **14**, 103–114
12. Gottlieb, B., Trifiro, M., Lumbroso, R., Vasiliou, D.M. & Pinsky, L. (1996) The androgen receptor gene mutations database. *Nucleic Acids Res.* **24**, 151–154
13. Matias, P.M., Donner, P., Coelho, R., Thomax, M., Peixto, C., Macedo, S., Otto, N., Joschko, S., Scholtz, P. & Wegg, A. (2000) Structural evidence for ligand specificity in the binding domain of the human androgen receptor. *J. Biol. Chem.* **275**, 26264–26171
14. Wu J.H. Gottlieb, B., Batist, G., Sulea, T., Purisima, E.O., Beitel, L.K. & Trifiro. M. (2003) Bridging structural biology and genetics by computational methods: an investigation into how the R774C mutation in the AR gene can result in complete androgen insensitivity syndrome. *Hum. Mutat.* **22**, 465–475
15. Gottlieb, B., Beitel, L.K. & Trifiro, M. (2001) Somatic mosaicism and variable expressivity. *Trends Genet.* **17**, 79–82
16. Debes, J.D. & Tindall, D.J. (2002) The role of androgens and the androgen receptor in prostate cancer. *Cancer Lett.* **187**, 1–7
17. Edwards, J., Krishma, N.S., Grigor, K.M. & Bartlett, J.M.S. (2003) Androgen receptor gene amplification and protein expression in hormone refractory prostate cancer. *Br. J. Cancer* **89**, 552–556
18. Culig, Z., Klocker, H., Bartsch, G. & Hobisch, A. (2002) Androgen receptors in prostate cancer. *Endocr. Relat. Cancer* **9**, 155–170

19. Culig, Z., Hobisch, A., Cronauer, M.V., Cato, A.C.B., Hittmair, A., Radmayr, C., Eberle, J., Bartsch, G. & Klocker, H. (1993) Mutant androgen receptor detected in an advanced-stage prostatic carcinoma is activated by adrenal androgens and progesterone. *Mol. Endocrinol.* **7**, 1541–1550
20. Taplin, M.-E., Bubley, G.J., Shuster, T.D., Frantz, M.E., Spooner, A.E., Ogata, G.K., Keer, H,N. & Balk, S.P. (1995) Mutation of the androgen-receptor gene in metastatic androgen-independent prostate cancer. *N. Engl. J. Med.* **332**, 1393–1398
21. Hyytinen, E.-J., Haapla, K., Thompson, J., Lappalainen, I., Roiha, M., Rantala, I., Helin, H.J., Janne, O.A., Vihinen, M., Plavimo, J.J. & Koivisto, P.A. (2002) Pattern of somatic androgen receptor gene mutations in patients with hormone-refractory prostate cancer. *Lab. Invest.* **82**, 1591–1592
22. Navarro, N., Luzardo, O.P., Fernandez, L., Chesa, N. & Diaz-Chico, B.N. (2002) Transition to androgen-independence in prostate cancer. *J. Steroid Biochem. Mol. Biol.* **81**, 191–201
23. Lobaccaro, J.M., Lumbroso, S., Belton, C., Galtier-Dereure, F., Bringer, J., Lesimple, T., Namer, M., Cutuli, B.F., Pujol, H. & Sultan, C. (1993) Androgen receptor gene mutation in male breast cancer. *Hum. Mol. Genet.* **2**, 1799–1802
24. La Spada, A.R. & Taylor, J.P. (2003) Polyglutamines placed in context. *Neuron* **36**, 681–684
25. Pinsky, L., Beitel, L.K. & Trifiro, M.A. (2001) Spinobulbar muscular atrophy. In The Metabolic and Molecular Basis of Inherited Disease, 8th edn (Scriver, C.R., Beaudet, A.L., Sly, W.S., Valle, D., Childs, B., Kinzler, K. & Vogelstein, B. eds.), pp. 4147–4157, McGraw-Hill, New York
26. Ferro, P., Catalano, M.G., Dell'Eva, R., Fortunati, N. & Pfeffer, U. (2002) The androgen receptor CAG repeat: a modifier of carcinogenesis? *Mol. Cell. Endocrinol.* **193**, 109–120
27. Rebbeck, T.R., Kantoff, P.W., Krithivas, K., Neuhausen, S., Blackwood, M.A., Godwin, A.K., Daly, M.B., Narod, S.A., Garber, J.E., Lynch, H.T. et al. (1999) Modification of the *BRCA1*-associated breast cancer risk by the polymorphic androgen-receptor CAG repeat. *Am. J. Hum. Genet.* **64**, 1371–1377
28. Dagan, E., Friedman, E., Paperna, T., Carmi, N. & Gershoni-Baruch, R. (2002) Androgen receptor CAG repeat length in Jewish Israeli women who are BRCA1/2 mutation carriers: association with breast/ovarian cancer phenotype. *Eur. J. Hum. Genet.* **10**, 724–728
29. Elhaji, Y.A., Gottlieb, B., Lumbroso, R., Beitel, L.K., Lumbroso, R. Foulkes, W.D., Pinsky, L. & Trifiro, M.A. (2001) The polymorphic CAG repeat of the androgen receptor gene: a potential role in breast cancer in woman over 40. *Breast Cancer Res. Treat.* **70**, 109–116
30. Lillie, E.O., Bernstein, L. & Ursin, G. (2003) The role of androgens and polymorphisms in the androgen receptor in the epidemiology of breast cancer. *Breast Cancer Res.* **5**, 167–173
31. Tut, T.G., Ghadessy, F.J., Trifiro, M.A., Pinsky, L. & Yong, E.L (1997) Long polyglutamine tracts in the androgen receptor are associated with reduced *trans*-activation, impaired sperm production, and male infertility. *J. Clin. Endocrinol. Metab.* **82**, 3777–3782
32. Casella, R, Madura, M.R., Misfud, A., Lipshultz, L.I., Yong, E.L. & Lamb, D.J. (2003) Androgen receptor gene polyglutamine length is associated with testicular histology in infertile patients. *J. Urol.* **169**, 224–227
33. Lund, A, Tapanianen, J.S., Lahdetie, J., Savontaus, M.L. & Aittomaki, K. (2003) Long CAG repeats in the AR gene are not associated with infertility in Finnish males. *Acta Obstet. Gynecol. Scand.* **82**, 162–166
34. Mhatre, A., Trifiro, M.A., Kaufman, M., Kazemi, E.P., Figlewicz, D., Rouleau. G. & Pinsky, L. (1993) Reduced transcriptional regulatory competence of the androgen receptor in X-linked spinal and bulbar muscular atrophy. *Nat. Genet.* **5**, 184–188
35. Irvine, R.A., Yu, M.C., Ross, R.K. & Coetzee, G.E. (1995) The CAG and GGC microsatellites of the androgen receptor gene are in linkage disequilibrium in men with prostate cancer. *Cancer Res.* **55**, 1937–1940
36. Zhang, L., Leeflang, E.P., Yu, J. & Arnheim, N. (1994) Studying human mutations by sperm typing: instability of CAG trinucleotide repeats in the androgen receptor gene. *Nat. Genet.* **7**, 531–535
37. Sasaki, M., Dahiya, R., Fujimoto, S., Ishikawa, M. & Oshimura, M. (2000) The expansion of the CAG repeat in exon 1 of the human androgen receptor gene is associated with uterine endometrial carcinoma. *Mol. Carcinog.* **27**, 237–244

10

Glucocorticoid and mineralocorticoid receptors and associated diseases

Tomoshige Kino[1] and George P. Chrousos

Pediatric and Reproductive Endocrinology Branch, National Institute of Child Health and Human Development, National Institutes of Health, Bethesda, MD 20892-1583, U.S.A.

Abstract

Adrenal corticosteroids, ie. glucocorticoids and mineralocorticoids, play important physiological roles in humans. Their actions are mediated by intracellular receptor molecules, the glucocorticoid receptor (GR) and mineralocorticoid receptor (MR), which function as hormone-dependent transcription factors. Ligand-activated receptors modulate the transcription rates of responsive genes by interacting with responsive elements in the promoters of these genes or by influencing the activities of other transcription factors, via protein–protein interactions. Natural inactivating mutations of the GR or MR genes have been reported in humans with significant clinical phenotypes. The former causes sporadic or familial glucocorticoid resistance characterized by generalized partial insensitivity of tissues to glucocorticoids and subsequent activation of the hypothalamic/pituitary/adrenal axis with resultant hyperandrogenism in children and women and/or mineralocorticoid excess symptoms in both sexes. The latter develop pseudohypoaldosteronism type 1, i.e. hypotension and hyperkalaemic acidosis, as a result of reduced aldosterone actions in the kidney. An activating mutation in the MR gene causing early-onset, periodic hypertension was reported recently. The biological relevance of the GR and MR receptors was also addressed in mice

[1]*To whom correspondence should be addressed (e-mail kinot@mail.nih.gov).*

whose GR or MR genes were inactivated or modified by gene targeting. The results were generally confirmatory of the concepts obtained by the human studies. Similarly, natural, compensated glucocorticoid and/or mineralocorticoid 'resistance' were described in several mammalian species, including non-human primates and rodents. Here we discuss the actions of GR and MR and the molecular defects of naturally occuring mutations in these receptors with associated pathophysiological changes.

Introduction

Two adrenal corticosteroids, the glucocorticoid cortisol and the mineralocorticoid aldosterone, exert profound influences on many physiological functions by virtue of their diverse roles in growth, development and maintenance of cardiovascular, metabolic and immune homoeostasis. Their actions are mediated by intracellular receptor proteins, the glucocorticoid receptor (GR) and mineralocorticoid receptor (MR), which function as hormone-activated transcription factors that regulate the expression of glucocorticoid and mineralocorticoid target genes, respectively [1].

The GR is expressed in almost all human tissues and organs. The presence of glucocorticoids is crucial for the integrity of central nervous system (CNS) function and for maintenance of cardiovascular, metabolic and immune homoeostasis. Increased glucocorticoid secretion during stress alters CNS function, assists with adjustments in energy expenditures and modulates the inflammatory/immune response. Since glucocorticoids possess a broad array of life-sustaining functions, only partial or incomplete glucocorticoid resistance — a state demonstrating reduced sensitivity/responsiveness to glucocorticoids — has been reported so far, suggesting that the complete inability of glucocorticoids to exert their effects on their target tissues is incompatible with human life. Over ten kindreds and individual patients suffering from congenital glucocorticoid resistance have been described to date, and the molecular mechanisms of their resistance have been analysed in some of them [2].

The MR mediates the sodium-retaining effects of aldosterone in the kidney, salivary glands, sweat glands and colon. In addition, the MR located in the CNS – also called corticosteroid type I receptor – appears to have a role in the regulation of the stress response and the feedback control of the hypothalamic/pituitary/adrenal axis (HPA axis) [2]. MR has a high affinity for both aldosterone and cortisol, and the circulating levels of cortisol are over 100 times higher than those of aldosterone. The MRs of the kidney distal convoluted tubule and possibly other mineralocorticoid target tissues are protected from the actions of cortisol by expression of 11β-hydroxysteroid dehydrogenase type 2, which converts cortisol into the inactive cortisone. Recently, inactivating mutations in the MR were shown to cause pseudohypoaldosteronism type 1 (PHA1), i.e. mineralocorticoid resistance [2a]. This disease, however, is mostly due to loss-of-function mutations in the subunits of the amiloride-sensitive sodium channel (ASSC), which represent a post-MR step in the signalling cascade of aldosterone in its target tissues.

Structure and actions of GRs and MRs

The GR and MR are members of the steroid/sterol/thyroid/retinoid/orphan receptor superfamily of nuclear transcription factors, with over 150 members currently cloned and characterized across species. Together with the progesterone, oestrogen and androgen receptors, GR and MR form the steroid receptor subfamily. Steroid receptors display a modular structure comprised of five or six regions (A–F), with the N-terminal A/B region harbouring an autonomous activation function (activation function 1) that catalyses transcriptional activity of the receptor by contacting and accumulating cofactors and basal transcriptional components, and the C and E regions corresponding to the DNA-binding domain (DBD) and ligand-binding domain (LBD) [2]. The human GR cDNA was isolated by expression cloning in 1985 [2b]. The genes of the GR consist of nine exons; its locus is on chromosome 5 (Figure 1A, a). It encodes two 3′ splice variants, GRα and GRβ, produced by alternative use of different terminal exons, 9α and 9β. The GRα encodes a 777-amino-acid protein, while the GRβ contains 742 amino acids. The first 727 amino acids from the N-terminus are identical in both isoforms. GRα possesses an additional 50 amino acids, while the GRβ encodes an additional 15 non-homologous amino acids in their C-terminus. GRα is the classic GR that binds to glucocorticoids and transactivates or transrepresses glucocorticoid-responsive promoters. On the other hand, human GRβ does not bind glucocorticoids and its physiological and pathological roles are not well known (Figure 1B) [1,2]. The cDNA for the human MR was isolated by low-stringency hybridization, using the human GR cDNA as a probe, in 1987 [2c]. The genes of the MR also consist of nine exons and its locus is on chromosome 4 (Figure 1A, b). Alternative 5′ promoters of the MR gene have been reported to regulate production of the same final receptor protein; the functional significance of this is not clear.

The GR and MR in their unliganded state are located primarily in the cytoplasm, as part of hetero-oligomeric complexes containing heat-shock proteins 90, 70 and 50, and possibly other proteins. After binding to its agonist ligand through their C-terminal LBD, the receptors undergo conformational changes, dissociate from the heat-shock proteins, dimerize and translocate into the nucleus through the nuclear pore via an active process. There, the ligand-activated GR and MR through their DBDs directly interact with DNA sequences in the promoter regions of target genes called glucocorticoid response elements (GREs). Both the GR and MR bind to and modulate transcription driven, for example, by the GRE-containing murine mammary tumour virus promoter [1]. Active endogenous GREs are present in the

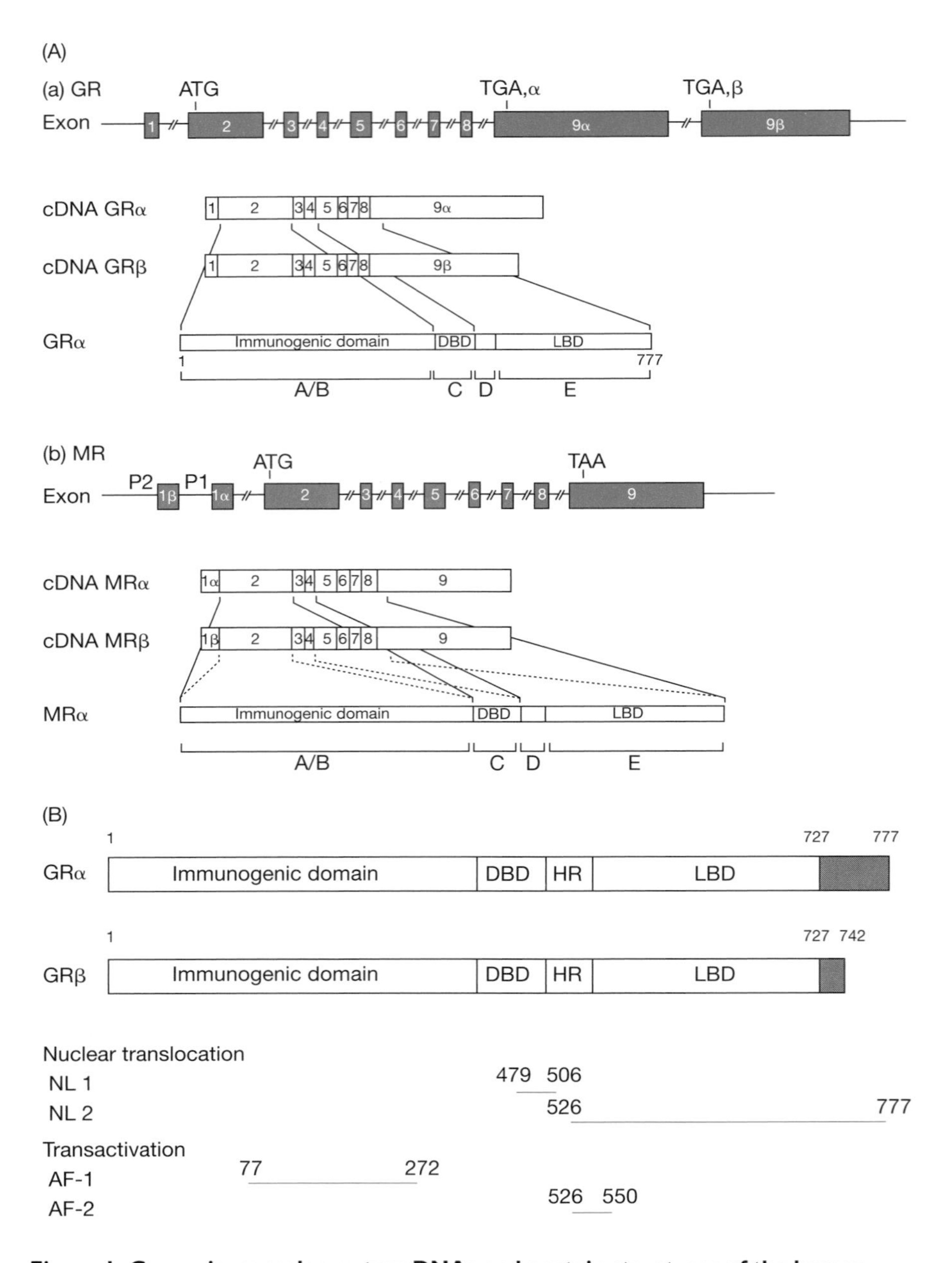

Figure 1. Genomic, complementary DNAs and protein structures of the human GR and MR.
(**A**) The human GR gene consists of ten exons. Exon 1 is an untranslated region, exon 2 codes for the immunogenic domain (A/B), exons 3 and 4 for the DBD (C), and exons 5–9 for the hinge region (D) and the LBD (E). The GR gene contains two terminal exons 9 (exons 9α and 9β), alternatively spliced to produce the classic GRα and the non-ligand-binding GRβ. (**B**) Functional domains of the human GR. The C-terminal grey-coloured domains in GRα and GRβ show the portions that are specific to each splice variant. HR, hinge region; NL 1 and 2, nuclear-localization signals 1 and 2; AF-1 and -2, activation functions 1 and 2. ATG, start codon; TGA and TAA, stop codons.

promoter regions of many glucocorticoid-responsive genes, whereas no specific mineralocorticoid-response elements have been characterized in the regulatory regions of genes physiologically regulated by aldosterone as yet. The GR as a dimer/monomer also modulates the transcription rates of non-GRE-containing genes regulated by other transcription factors, such as activator protein 1, nuclear factor κB and signal transducer and activator of transcription 5 (STAT5), through protein–protein interactions with these factors.

The promoter-bound GR and MR stimulate the transcription rates of responsive genes by facilitating the formation of the transcription-initiation complex, including the RNA polymerase II and its ancillary factors. In addition to these molecules, GR and MR, via their two transactivation domains attract several proteins and protein complexes, so-called co-activators, that help transmit the glucocorticoid complex signal to the transcription-initiation complex as well as contain intrinsic histone acetyltransferase activity, through which they loosen the chromatin structure and facilitate access and/or binding of transcription machinery components to DNA [3]. They include the homologous p300 and cAMP-response element-binding protein (CREB)-binding protein (CBP), the p160 steroid receptor co-activators and the p300/CBP-associated factor (pCAF). The p300/CBP co-activators may serve as macromolecular docking 'platforms' for transcription factors from several signal transduction cascades, including, in addition to nuclear receptors, CREB, activator protein 1, nuclear factor κB, p53, Ras-dependent growth factor and STATs. pCAF, originally reported as a human homologue of yeast Gcn5 that interacts with p300/CBP, is also a broad co-activator with histone acetyltransferase activity. Steroid receptors preferentially interact with the p160 family of co-activators: steroid receptor co-activator-1 (SRC-1), transcription intermediate factor-II (TIF-II) or GR-interacting protein-1 (GRIP-1), also called SRC-2, the p300/CBP/co-integrator-associated protein (p/CIP), activator of thyroid receptor (ACTR) or receptor-associated co-activator-3 (RAC3), also called SRC-3.

The p300/CBP and p160 family co-activators contain one or more copies of the co-activator signature motif sequence Leu-Xaa-Xaa-Leu-Leu (LXXLL), also called nuclear-receptor-binding box (NRB), through which they directly bind the GR and MR [3]. It appears that p160 co-activators are first attracted to the DNA-bound steroid receptors, where they help accumulate p300/CBP and pCAF to the promoter region through their mutual interactions, indicating that p160 proteins play a central role in the transactivation by steroid receptors.

In addition to the co-activator molecules, there are several proteins whose function is to retain the steroid receptors in the repressed, inactive state. These molecules, called co-repressors, may have deacetylase activity themselves or may attract other molecules with deacetylase activity, through which they tighten chromatin structure and prevent transcription machinery components from binding to DNA [3]. Co-repressors also employ a sequence close to the LXXLL motif to bind to nuclear receptors. The distinction between co-activators and co-repressors is not absolute, as there are now several examples of co-activators acting as co-repressors, and vice versa, depending on cell or tissue type, or state of cell activation [4].

The recent development of GR- and MR-knockout mice has provided new insights into the biological activities of these receptors. Mice harbouring complete inactivation of the GR gene died at birth from severe respiratory distress syndrome due to a deficit of lung surfactant [5]. In these mice, transcription of genes encoding gluconeogenic enzymes in the liver was decreased, proliferation of erythroid progenitors was impaired and the HPA axis was strongly up-regulated, indicating that GR plays an important role in the regulation of these activities. In contrast with GR-knockout mice, GR-knockin mice, with a mutated GR defective in GRE-mediated transactivation but intact in protein–protein interaction with other transcription factors, survived and procreated [5]. These animals demonstrated defects in the induction of gluconeogenic enzymes and proliferation of erythroid progenitors, while most of the immune-suppressive effects of glucocorticoids were preserved. Regarding their HPA axis, suppression of corticotropin-releasing hormone (CRH) synthesis by glucocorticoids was maintained, whereas pro-opiomelanocortin expression was up-regulated.

By using the Cre-LoxP-mediated recombination reaction, mice harbouring a brain-specific GR-knockout were also developed [6]. These mice demonstrated severe impairment of their HPA axis, resulting in increased corticosterone levels due to a loss of its negative-feedback action on the secretion of CRH and corticotropin (adrenocorticotrophic hormone, 'ACTH'). These animals developed symptoms that mimicked the phenotype of the glucocorticoid-excess (Cushing) syndrome. In addition, they had impaired behavioural responses to external stressors and displayed reduced anxiety, indicating that the brain GR plays roles in emotional behaviour and cognitive functions of the brain.

MR-deficient mice died in the second week after birth, because of high renal salt wasting and hyponatraemia, hyperkalaemia and acidosis. They had very high plasma renin activity and increased plasma aldosterone levels. This phenotype of the MR-knockout mice is similar to that of PHA1 syndrome in humans [7]. The activity of the ASSC, whose defect is known to be a major cause of human PHA1, was almost undetectable in their kidneys, while the mRNA levels of the three ASSC subunits were preserved, indicating that MR may affect translational or post-translational steps in the expression of functional ASSC proteins.

The MR functions as a GR in the brain, since it is exposed to relatively high concentrations of glucocorticoids due to no expression of 11β-hydroxysteroid dehydrogenase type 2, which in the kidney converts active cortisol/corticosterone into inactive cortisone. MR-knockout mice, whose electrolyte deficits were corrected by exogenous NaCl administration, demonstrated a decreased number of hippocampal granular cells and decreased neurogenesis, while brain-specific GR-knockout mice had a normal expression of hippocampal granular cells and neurogenesis [8]. These MR-mediated brain functions may contribute to the development of pathological changes in the hippocampus that appear with normal aging, in subjects exposed to chronic stress, and in patients with affective disorders.

Natural physiological steroid hormone 'resistance' in animals

New World monkeys

Some animal species demonstrate generalized (whole-body) insensitivity to several steroid hormones. New World monkeys (infraorder Platyrrhini, superfamily Ceboidea), such as the owl (*Aotus*), titi (*Callicebus*) and squirrel (*Saimiri*) monkeys, have elevated levels of cortisol, which compensate for the low binding affinity of their GRs to glucocorticoids [9,10]. These animals also demonstrate increased levels of other steroid hormones, such as progesterone, oestrogen, testosterone and aldosterone, indicating that they have pan-steroid resistance, possibly via a common mechanism shared by these affected receptors [10]. Since squirrel monkey GR in cultured cell lines has only a mildly decreased affinity to glucocorticoids *in vitro* and since the cytoplasmic fraction from squirrel monkey cells reduces ligand-binding activity of human GR, it has been suggested that some unknown cytosolic factor(s) might regulate ligand-binding potency of GR in affected New World monkeys.

Recently, FK506-binding protein (FKBP) 51, an immunophilin that inhibits the association of GR with heat-shock protein 90 and thus reduces the affinity of GR to ligands, was postulated to cause glucocorticoid insensitivity in New World monkeys [11]. Indeed, immunophilin FKBP51 is expressed at large amounts in all affected New World monkeys tested and administration of FK506, which binds and activates FKBP51, reverses the glucocorticoid-insensitivity state of squirrel monkey cells *in vitro*.

Guinea pigs

Guinea pigs also demonstrate generalized glucocorticoid 'resistance' associated with high levels of glucocorticoids [12]. Similarly to New World monkeys, their GR shows low affinity to ligands, but the cause of such a change may be due to the sequence alteration of the receptor itself. Indeed, the LBD of guinea pig GR has five amino acid substitutions, which are preserved in many other species. The alteration of these amino acids reduces the affinity of the guinea pig GR to glucocorticoids, possibly by inducing a conformational change in the ligand-binding pocket.

Prairie voles

Prairie voles (*Microtus ochrogaster*) also demonstrate extremely high plasma glucocorticoid concentrations in the absence of any apparent negative causes of glucocorticoid excess [13]. They have a significantly higher adrenal/body-weight ratio, 5–10-fold greater basal plasma corticosterone and 2–3-fold greater basal plasma corticotropin concentrations than montane voles (*Microtus montanus*) and rats. Their plasma corticosterone levels are responsive to both stress and circadian cues but are resistant to the administration of the synthetic glucocorticoid, dexamethasone. In agreement with these physiological results, GR in their brain and liver shows significantly lower affinity to glucocorticoids compared with that in the same tissues of rats. These pieces of evidence indicate that resistance to glucocorticoid hormones in peripheral tissues, possibly due to lower affinity of their GR to glucocorticoids, may account for their high glucocorticoid levels.

Pathological changes of GR and MR activities in humans

Humans also develop generalized insensitivity/hypersensitivity to single or several steroid hormones that are caused by mutations in steroid hormone receptors. Below, we explain the genetic and biological changes seen in patients with GR or MR mutations.

GR mutations

Familial/sporadic glucocorticoid resistance syndrome was first described in 1976, as a disorder characterized by hypercorticosolism without Cushingoid features. Since then, over ten kindreds and sporadic cases with abnormalities of the number of binding sites for glucocorticoids, affinity for glucocorticoids, stability and translocation into the nucleus have been reported [2,14]. However, to date, the molecular defects have been elucidated in five kindreds and three sporadic cases (Figure 2, Table 1). The index propositus, a patient through whom the particular mutation(s) is/are found in his or her family, of the original kindred was a homozygote for a single non-conservative point mutation, replacing aspartic acid with valine at amino acid 641 in GR LBD; this mutation reduced binding affinity for dexamethasone by 3-fold and caused a concomitant loss of transactivation activity [15].

The proposita of the second family had four-base deletion at the 3′ boundary of exon 6, removing a donor splice site [16]. This resulted in complete ablation of one of the GR alleles in affected members of the family. The propositus of the third kindred had a single homozygotic point mutation at amino acid 729 (valine to isoleucine) in LBD, which reduced both the affinity and transactivation activity of the GR [17]. There was also an interesting sporadic case of a man with a *de novo* germline heterozygotic GR mutation at amino acid 559 (isoleucine to asparagine) also in LBD close to nuclear-localization signal 1. This mutant GR bound no ligand but exerted dominant-negative activity on the wild-type receptor by preventing or retarding the translocation of the wild-type receptor into the nucleus [18,19].

Study of a fifth case/kindred with glucocorticoid resistance and a heterozygotic GR mutation in the LBD (amino acid 747, replacing isoleucine

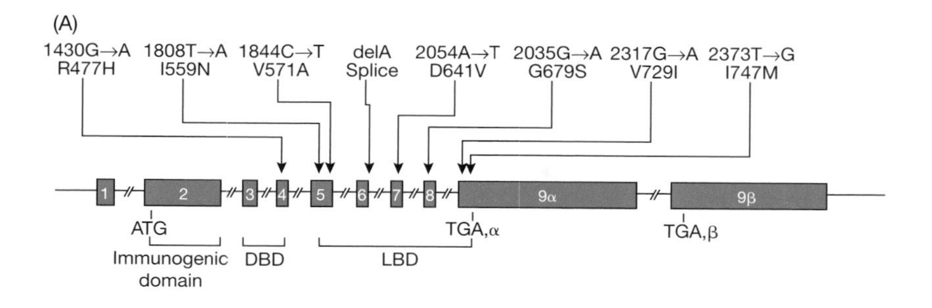

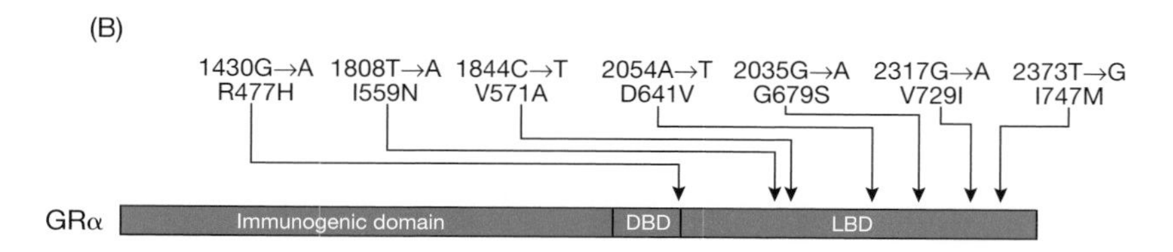

Figure 2. Location of the known mutations of the GR in its genomic (A) and protein (B) structures
ATG, start codon; TGA, stop codon.

with methionine) was recently completed [20]; the mutant receptor had mildly reduced affinity for dexamethasone and markedly decreased transactivation activity; interestingly, it also had dominant-negative activity on the wild-type receptor. The mutation was located just a few amino acids before the helix 12 of the GR LBD. This mutant receptor could not bind to the nuclear co-activator signature motif LXXLL of the p160 type nuclear receptor co-activator but still associated with this co-activator through its intact activation function 1 domain. Overexpression of p160 co-activator diminished the dominant-negative activity of the mutant receptor, suggesting that defective interaction of the mutant receptor with p160 co-activator might explain its dominant-negative activity on the wild-type receptor.

The sixth and seventh sporadic cases were also found as heterozygous mutations with histidine replacing arginine at amino acid 477, and a glycine-to-serine change at amino acid 697, respectively [21]. The former is located in the second zinc finger in the DBD. This mutant receptor has no transactivation activity due to impaired binding to GREs. The latter mutation is located in the LBD, outside of the ligand-binding pocket. This mutation caused 50% reduction of ligand-binding affinity with comparable reduction of the transactivation activity. Since these two mutant receptors were found in the heterozygotic condition, they might also behave as dominant-negative mutants to the wild-type receptor, suppressing its activity.

The proposita of the eighth case, a girl born with ambiguous genitalia, had a homozygotic point mutation replacing valine with alanine at amino acid 571

Table 1. Pathological mutations in the GR gene

Position of mutation				
cDNA	**Amino acid**	**Biochemical phenotype**	**Genotype/transmission**	**Reference**
A-2054→T	Asp-641→Val	Affinity ↓ Transactivation ↓	Homozygote/autosomal recessive	[15]
Δ4 at the 39 boundary of exon and intron 6		GR number ↓ Inactivation of the affected allele	Heterozygote/autosomal dominant	[16]
G-2317→A	Val-729→Ile	Affinity ↓ Transactivation ↓	Homozygote/autosomal recessive	[17]
T-1808→A	Ile-559→Asn	Number ↓ Transactivation ↓ Dominant-negative activity	Heterozygote/sporadic	[18,19]
T-2373→G	Ile-747→Met	Affinity ↓ Transactivation ↓↓ Dominant-negative activity	Heterozygote/autosomal dominant	[20]
G-1430→A	Arg-477→His	Transactivation (−)	Heterozygote/sporadic	[21]
G-2035→A	Gly-679→Ser	Affinity ↓ Transactivation ↓	Heterozygote/sporadic	[21]
A-1844→T	Thr-571→Cys	Affinity ↓ Transactivation	Homozygote/autosomal recessive	[22]

in the LBD [22]. The mutant receptor had a 6-fold reduction in its binding affinity for dexamethasone and 10–50-fold lower transactivation activity than the wild-type receptor. Interestingly, the proposita was also a carrier of 21-hydroxylase deficiency, suggesting that association with this congenital disorder exacerbated the hyperandrogenism and virilization potential of the glucocorticoid-resistance syndrome.

Pathophysiology of glucocorticoid-resistance syndrome

A complex negative-feedback system exists in the human CNS that regulates glucocorticoid homoeostasis. The regulatory circuit for glucocorticoid secretion in the CNS detects and integrates external signals through numerous parts of the CNS, and such inputs are transduced to the paraventricular nucleus of the hypothalamus, which produces CRH, the major stimulator of corticotropin from the anterior lobe of the pituitary gland, and arginine vasopressin (AVP). Axons from this nucleus project to the median eminence and secrete CRH into the hypophyseal portal blood system, which then circulates to the anterior pituitary gland and stimulate corticotropin production and secretion. Glucocorticoids exert negative-feedback effects on both hypothalamic CRH and AVP secretion and inhibit pituitary corticotropin secretion itself. In addition, glucocorticoids influence the activity of suprahypothalamic centres, including the paraventricular nucleus, that control the activity of CRH and AVP neurons [2].

This complex regulatory system is adjusted to higher levels in patients with loss-of-function GR mutations, since the mutated GRs need more glucocorticoids to exert normal biological effect, including suppression of the regulatory system. Thus the mutations result in compensatory increases in corticotropin and cortisol secretion (Figure 3). The patients retain the circadian rhythm and responsiveness of cortisol to stress and are resistant to single or multiple doses of dexamethasone. Although adequate compensation is apparently achieved by elevated cortisol concentrations in the great majority of the patients described, excess corticotropin secretion also results in increased production of adrenal steroids with mineralocorticoid activity and enhanced secretion of adrenal androgens. The former, together with cortisol, is responsible for causing symptoms and signs of mineralocorticoid excess, such as hypertension and/or hypokalaemic alkalosis, which is caused by compensatory increase of the H^+ secretion from the renal tubules in an attempt to recover secreted K^+ from the filtrated fluid due to mineralocorticoid excess. On the other hand, the latter causes varying manifestations of hyperandrogenism, such as acne, hirsutism, male-pattern baldness, menstrual irregularities and infertility in women. Precocious puberty has been seen in a boy due to early and excessive prepubertal adrenal androgen secretion and ambiguous genitalia in a genetic female due to excessive production of adrenal androgens by the fetal adrenal zone. In the male, oligospermia and infertility have been observed, possibly as a result of disturbances in follicle-stimulating hormone regulation

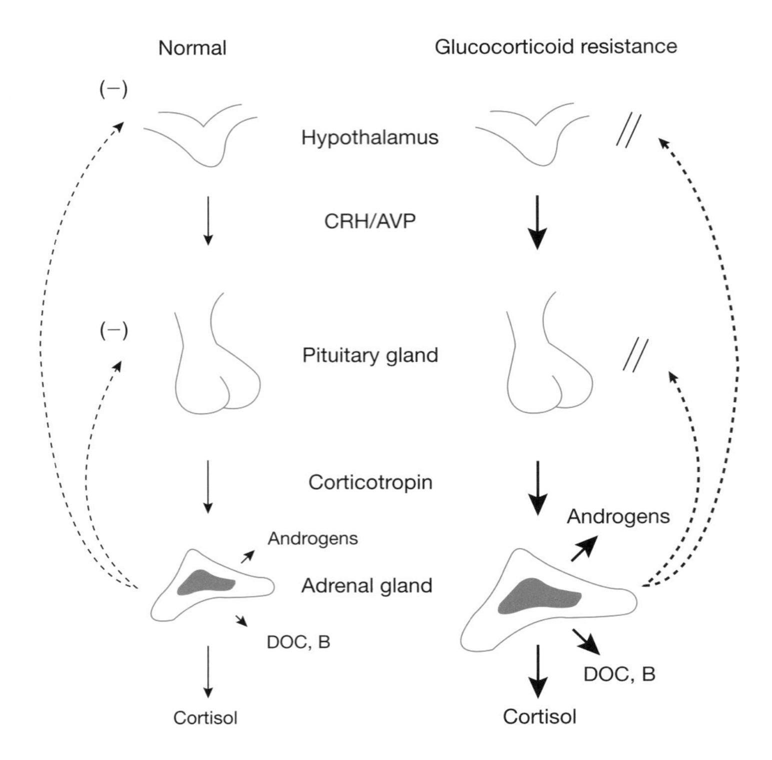

Figure 3. Pathophysiological mechanism of glucocorticoid resistance induced by loss-of-function GR mutations
The elaborate negative-feedback mechanism responsible for maintenance of glucocorticoid homoeostasis compensates for tissue insensitivity to glucocorticoids due to GR mutations by resetting the HPA axis at a higher level. Thus CRH/AVP, corticotropin and cortisol secretion are increased. The compensatory increase in corticotropin production augments the secretion of cortisol and glucocorticoid precursors with mineralocorticoid activity, as well as the secretion of several adrenal androgens, including Δ4-androstenedione, which has considerable androgen activity. DOC, deoxycorticosterone; B, cortisosterone.

caused by excessive adrenal androgens. However, the spectrum of clinical manifestations in patients with GR mutations is broad, as a large number of subjects are asymptomatic and show only biochemical changes.

Treatment of glucocorticoid-resistance syndrome

Patients are treated with high doses of synthetic glucocorticoids with low mineralocorticoid activity. The goal is to suppress the increased levels of corticotropin, which cause overproduction of mineralocorticoids and androgens (Figure 3) [2]. As all cases described to date have had partial inactivation of GR activity, synthetic potent glucocorticoids (e.g. dexamethasone) in minimal intrinsic mineralocorticoid activity is a rational approach. These steroids achieve activation of the mutated GR in homozygous cases or of the wild-type receptor in heterozygous cases sufficient to suppress the compensatory increases of corticotropin, and hence the production of the adrenal mineralocorticoids and androgens causing the clinical manifestations of the condition. The patients should be treated with high, individualized doses of oral dexamethasone (1–3 mg/day). Dexamethasone indeed suppresses corticotropin and therefore endogenous cortisol, deoxycorticosterone, corticosterone and adrenal androgen secretion, correcting the mineralocorticoid and androgen excess states of these patients.

MR mutations

Inactivating mutation of the MR

The mechanism whereby aldosterone stimulates sodium transport in its target tissues may involve the synthesis of a protein associated with the function of the ASSC. The latter is located in the apical membrane of epithelial cells of the renal distal convoluted tubule, and in the plasma membranes of cells in other tissues involved with salt conservation. The phenotype of patients with loss-of-function mutations of the MR mimics that of patients with defects in the subunits of the ASSC who represent the bulk of patients with PHA1 [23,24].

Cheek and Perry first reported PHA1 in an infant with severe salt-wasting syndrome in 1958; PHA1 was subsequently reported in more than 70 patients [25]. This syndrome usually presents in infancy with urinary salt wasting and failure to thrive. The levels of plasma renin activity and aldosterone concentrations are markedly elevated. Approx. one-fifth of these cases are familial. All patients have renal tubular unresponsiveness to aldosterone, whereas some have multiple mineralocorticoid target tissue involvement, including the sweat and salivary glands and the colonic epithelium.

In kindreds with PHA1, both an autosomal-dominant and recessive form of genetic transmission have been observed. The autosomal-recessive form was associated with severe disease, with manifestations persisting into adulthood. We and others failed to find pathological mutations in the MR gene in our sporadic and familial cases with autosomal-recessive PHA1, and concluded that, most probably, this condition was due to a defect in a post-MR step of aldosterone action [26–28]. Indeed, in 1996, PHA1 was found to be caused by loss-of-function mutations in genes encoding subunits of the ASSC [23,24]. However, Geller et al. [2a] identified heterozygotic MR gene loss-of-function mutations in one sporadic case and four autosomal-dominant cases of PHA1 (Figure 4, Table 2) [2a]. These included two frameshift mutations, each deleting a single base pair in exon 2; the resultant frame shifts led into a gene product lacking the entire DNA- and hormone-binding domains, as well as a dimerization motif. Two families had an identical mutation, introducing a premature termination codon in exon 2 at position 537. One case showed a single-base-pair deletion in the intron 5 splice donor site. Subsequently, Tajima et al. [29] reported the fifth family of the pathological mutation in the MR gene from the patients with the autosomal-dominant PHA1 [29]. The propositus had a single heterozygotic point mutation at amino acid 924 (leucine to proline) in LBD of MR. The mutation

(A)

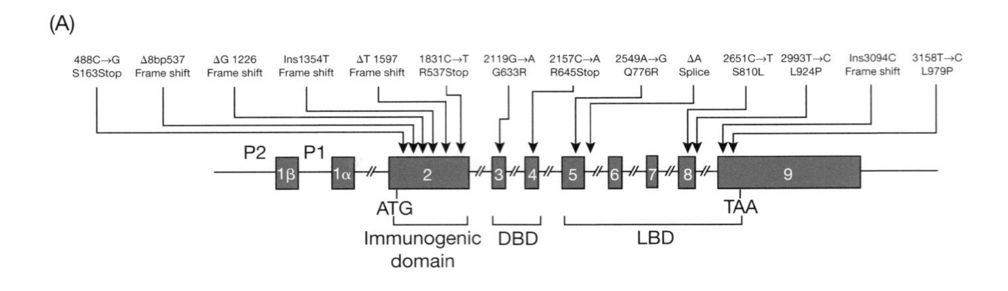

(B)

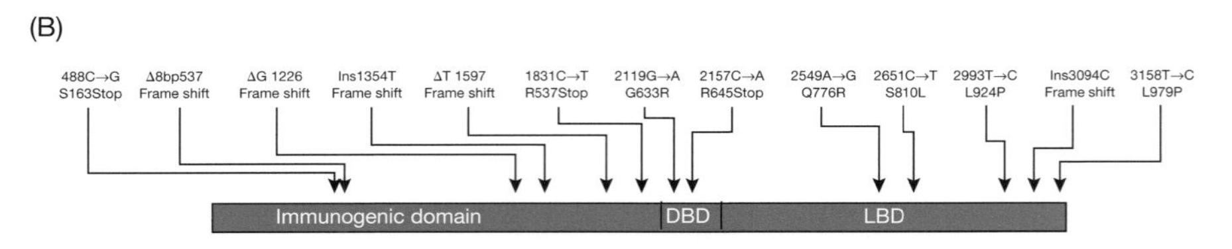

Figure 4. Location of the known mutations of the MR in its genomic (A) and protein (B) structures
ATG, start codon; TAA, stop codon.

completely abolished the transactivation activity of the mutant MR on the murine mammary tumour virus promoter. Since the mutation is located in helix 11 of the MR LBD that forms the ligand-binding pocket with helices 3, 4 and 12, the mutant receptor might lose transcriptional activity through its inability to bind ligands. The sixth inactivating MR mutation was found as a sporadic PHA1, harbouring a heterozygous frame-shift mutation that inserted a cytosine at position 3094 in exon 9 that resulted in a nonsense protein from 958 and a first stop codon at position 1012 [30]. The seventh pathological mutation of the MR gene was found in a German family as a heterozygotic mutation in exon 2 replacing serine at amino acid 163 to stop codon (Ser-163→stop) [31]. The propositus demonstrated a clear phenotype of autosomal-dominant PHA1, while his father had no clinical signs of PHA1 in his entire life, indicating that phenotype of the MR mutation is heterologous even within a family, possibly due to their genetic background and/or developmental influences. Recently, Sartorato et al. [32] reported a large study analysing 14 families with an autosomal-dominant PHA1 phenotype and found six hetrozygotic mutations. They reported two frame-shift mutations in exon 2 (insertion of Thr-1354, deletion of 8 bp at position 537) and one nonsense mutation in exon 4 (Cys-2157→Ala, Cys-645→stop), which produce several truncated MR molecules devoid of the entire LBD. They also reported three missense mutations (Gly-633→Arg, Gln-776→Arg and Leu-979→Pro); the Gly-633→Arg mutation, situated in the DBD, demonstrated attenuated transactivation activity possibly due to reduced binding activity of the mutant receptor to DNA. The Gln-776→Arg and Leu-979→Pro mutations, also in the LBD, had reduced or absent binding activity to

Table 2. Pathological mutations in the MR gene

Position of mutation				
cDNA	**Amino acid**	**Genotype/transmission**	**Phenotype**	**Reference**
ΔG-1226	Frame shift	Heterozygote	PHA1	[2a]
ΔT-1597	Frame shift	Heterozygote	PHA1	[2a]
C-1831→T	Arg-537→stop	Heterozygote	PHA1	[2a]
ΔA at the 3′ boundary of exon and intron 5		Heterozygote	PHA1	[2a]
C-2651→T	Ser-810→Leu	Heterozygote	Hypertension	[33]
T-2993→C	Leu-924→Pro	Heterozygote	PHA1	[29]
Insertion of C-3094	Frame shift	Heterozygote	PHA1	[30]
C-488→T	Ser-163→stop	Heterozygote	PHA1	[31]
Δ8 bp at 537	Frame shift	Heterozygote	PHA1	[32]
Insertion of 1357T	Frame shift	Heterozygote	PHA1	[32]
G-2119→A	Gly-633→Arg	Heterozygote	PHA1	[32]
C-2157→A	Arg-645→stop	Heterozygote	PHA1	[32]
A-2549→G	Gln-776→Arg	Heterozygote	PHA1	[32]
CT-3158→C	Leu-979→Pro	Heterozygote	PHA1	[32]

aldosterone and corresponding blunted or absent transactivation capacity. Leu-979→Pro also functioned as a transdominant-negative mutant to the wild-type MR on its transactivation of responsive genes.

As indicated, all reported MR mutations causing PHA1 phenotype were found in the heterozygote state, i.e. in only one of the two alleles of the MR gene. They consisted of a frameshift, insertion of a premature termination codon or a point missense mutation, indicating that PHA1 phenotype can be caused by haploinsufficiency, a condition harbouring the heterozygotic mutation. The patients developed PHA1 in an early stage of their life and their clinical condition improved with age, although plasma aldosterone concentrations and renin activity remained high. These clinical findings suggest that the intact MR gene is required for the reabsorption of salt at least in infancy, and some other yet unknown mechanisms may overcome defects of MR function in older ages. Further work is necessary to address this issue.

Treatment of PHA1 is done by the supplementation of patients with NaCl. The amounts of salt may be different from patient to patient depending on their degree of salt wasting. The salt-administration requirements may decrease with advancing age.

Activating mutation of the MR

The first activating MR mutation was found in a patient with early-onset hypertension that was markedly exacerbated in pregnancy (Figure 4, Table 2) [33]. The propositus had a heterozygotic point mutation, Ser-810→Leu. This mutation is localized in helix 5 of the MR LBD: the leucine side chain projects into the ligand-binding pocket, potentially forming additional van der Waals interactions with Ala-773 of helix 3 and the carbon-19 methyl group of steroid hormones. Therefore, this mutation may change the ligand specificity of the mutant receptor, conferring increased binding to progesterone in addition to mineralocorticoids and glucocorticoids. Thus a patient with such a mutation may have worsening hypertension in pregnancy due to the activation of the mutant receptor by progesterone, secreted from the placenta or by the physiologically increased levels of cortisol in pregnancy. In addition, it was recently reported that cortisone and 11-dehydrocorticosterone bind to the mutant receptor with high affinity and may thus cause early-onset hypertension in affected men and non-pregnant women [34].

Conclusions

We have described the molecular defects observed in the GR and MR genes, the pathophysiological mechanisms resulting in disease and have suggested rational therapeutic interventions. Although these mutations are rare, they provide strong insight into the physiological importance of hormonal actions of glucocorticoids and mineralocorticoids, and may provide clues to unknown important functions of these hormones.

Summary

- *Adrenal corticosteroids, i.e. glucocorticoids and mineralocorticoids, play important roles in human physiology. The former are necessary for the maintenance of CNS function and cardiovascular, metabolic and immune homoeostasis, while the latter play a critical role in the retention of salt in the kidney, salivary glands, sweat glands, and colon.*
- *The actions of these hormones are mediated by intracellular receptor molecules, the GR and MR, which function as hormone-dependent transcription factors. Ligand-activated receptors modulate the transcription rates of responsive genes by interacting with responsive elements in the promoters of these genes and/or by influencing the activities of other transcription factors, via protein–protein interactions. The biological activities of these receptors were examined recently in animals and human patients whose genes for these receptors are influenced by genetic modifications.*
- *Natural physiological steroid hormone 'resistance' in animals has been reported in New World monkeys, Guinea pigs and prairie voles, with distinct mechanisms affecting the biological actions of several steroid hormone receptors.*
- *In human patients, the familial/sporadic glucocorticoid-resistance syndrome is characterized by partial insensitivity to glucocorticoids with concomitant hypercorticolism, but without Cushingoid features. This syndrome is caused by loss-of-function mutations of the GR gene and is associated with hyperandrogenism and/or hypermineralocorticoidism.*
- *In human patients, inactivating mutations of the MR cause pseudohypoalsosteronism type 1, which presents in infancy with urinary salt wasting and failure to thrive, in spite of high levels of circulating aldosterone. Recently, an activating mutation of the MR was reported as early-onset hypertension that was markedly exacerbated in pregnancy. This mutation changed the ligand specificity of the receptor, conferring increased binding to progesterone in addition to aldosterone and cortisol.*

References

1. Bamberger, C.M., Schulte, H.M. & Chrousos, G.P. (1996) Molecular determinants of glucocorticoid receptor function and tissue sensitivity to glucocorticoids. *Endocr. Rev.* **17**, 245–261
2. Kino, T., Vottero, A., Charmandari, E. & Chrousos, G.P. (2002) Familial/sporadic glucocorticoid resistance syndrome and hypertension. *Ann. N.Y. Acad. Sci.* **970**, 101–111

2a. Geller, D.S., Rodriguez-Soriano, J., Vallo Boado, A., Schifter, S., Bayer, M., Chang, S.S. & Lifton, R.P. (1998) Mutations in the mineralocorticoid receptor gene cause autosomal dominant pseudohypoaldosteronism type I. *Nat. Genet.* **19**, 279–281

2b. Hollenberg, S.M., Weinberger, C., Ong, E.S., Cerelli, G., Oro, A., Lebo, R., Thompson, E.B., Rosenfeld, M.G., & Evans, R.M. (1985) Primary structure and expression of a functional human glucocorticoid receptor cDNA. *Nature (London)* **318**, 635–641

2c. Arriza, J.L., Weinberger, C., Cerelli, G., Glaser, T.M., Handelin, B.L., Hausman, D.E. & Evans, R.M. (1987) Cloning of human mineralocorticoid receptor complimentary DNA: structural and functional kinship with the glucocorticoid receptor. *Science* **237**, 268–275

3. McKenna, N.J., Lanz, R.B. & O'Malley, B.W. (1999) Nuclear receptor coregulators: cellular and molecular biology. *Endocr. Rev.* **20**, 321–344

4. Kino, T., Nordeen, S.K. & Chrousos, G.P. (1999) Conditional modulation of glucocorticoid receptor activities by CREB-binding protein (CBP) and p300. *J. Steroid Biochem. Mol. Biol.* **70**, 15–25

5. Kellendonk, C., Tronche, F., Reichardt, H.M. & Schutz, G. (1999) Mutagenesis of the glucocorticoid receptor in mice. *J. Steroid Biochem. Mol. Biol.* **69**, 253–259

6. Tronche, F., Kellendonk, C., Kretz, O., Gass, P., Anlag, K., Orban, P.C., Bock, R., Klein, R. & Schutz, G. (1999) Disruption of the glucocorticoid receptor gene in the nervous system results in reduced anxiety. *Nat. Genet.* **23**, 99–103

7. Berger, S., Bleich, M., Schmid, W., Cole, T.J., Peters, J., Watanabe, H., Kriz, W., Warth, R., Greger, R. & Schutz, G. (1998) Mineralocorticoid receptor knockout mice: pathophysiology of Na^+ metabolism. *Proc. Natl. Acad. Sci. U.S.A.* **95**, 9424–9429

8. Gass, P., Kretz, O., Wolfer, D.P., Berger, S., Tronche, F., Reichardt, H.M., Kellendonk, C., Lipp, H.P., Schmid, W. & Schutz, G. (2000) Genetic disruption of mineralocorticoid receptor leads to impaired neurogenesis and granule cell degeneration in the hippocampus of adult mice. *EMBO Rep.* **1**, 447–451

9. Chrousos, G.P., Renquist, D., Brandon, D., Eil, C., Pugeat, M., Vigersky, R., Cutler, G.B., Jr, Loriaux, D.L. & Lipsett, M.B. (1982) Glucocorticoid hormone resistance during primate evolution: receptor-mediated mechanisms. *Proc. Natl. Acad. Sci. U.S.A.* **79**, 2036–2040

10. Chrousos, G.P., Loriaux, D.L. & Lipsett, M.B. (1986) Steroid Hormone Resistance: Mechanisms and Clinical Aspects, Plenum Press, New York

11. Scammell, J.G., Denny, W.B., Valentine, D.L. & Smith, D.F. (2001) Overexpression of the FK506-binding immunophilin FKBP51 is the common cause of glucocorticoid resistance in three New World primates. *Gen. Comp. Endocrinol.* **124**, 152–165

12. Keightley, M.C., Curtis, A.J., Chu, S. & Fuller, P.J. (1998) Structural determinants of cortisol resistance in the guinea pig glucocorticoid receptor. *Endocrinology* **139**, 2479–2485

13. Taymans, S.E., DeVries, A.C., DeVries, M.B., Nelson, R.J., Friedman, T.C., Castro, M., Detera-Wadleigh, S., Carter, C.S. & Chrousos, G.P. (1997) The hypothalamic-pituitary-adrenal axis of prairie voles (*Microtus ochrogaster*): evidence for target tissue glucocorticoid resistance. *Gen. Comp. Endocrinol.* **106**, 48–61

14. Kino, T. & Chrousos, G.P. (2001) Glucocorticoid and mineralocorticoid resistance/hypersensitivity syndromes. *J. Endocrinol.* **169**, 437–445

15. Hurley, D.M., Accili, D., Stratakis, C.A., Karl, M., Vamvakopoulos, N., Rorer, E., Constantine, K., Taylor, S.I. & Chrousos, G.P. (1991) Point mutation causing a single amino acid substitution in the hormone binding domain of the glucocorticoid receptor in familial glucocorticoid resistance. *J. Clin. Invest.* **87**, 680–686

16. Karl, M., Lamberts, S.W., Detera-Wadleigh, S.D., Encio, I.J., Stratakis, C.A., Hurley, D.M., Accili, D. & Chrousos, G.P. (1993) Familial glucocorticoid resistance caused by a splice site deletion in the human glucocorticoid receptor gene. *J. Clin. Endocrinol. Metab.* **76**, 683–689

17. Malchoff, D.M., Brufsky, A., Reardon, G., McDermott, P., Javier, E.C., Bergh, C.H., Rowe, D. & Malchoff, C.D. (1993) A mutation of the glucocorticoid receptor in primary cortisol resistance. *J. Clin. Invest.* **91**, 1918–1925

18. Karl, M., Lamberts, S.W., Koper, J.W., Katz, D.A., Huizenga, N.E., Kino, T., Haddad, B.R., Hughes, M.R. & Chrousos, G.P. (1996) Cushing's disease preceded by generalized glucocorticoid resistance: clinical consequences of a novel, dominant-negative glucocorticoid receptor mutation. *Proc. Assoc. Am. Physicians* **108**, 296–307

19. Kino, T., Stauber, R.H., Resau, J.H., Pavlakis, G.N. & Chrousos, G.P. (2001) Pathologic human GR mutant has a transdominant negative effect on the wild-type GR by inhibiting its translocation into

the nucleus: importance of the ligand-binding domain for intracellular GR trafficking. *J. Clin. Endocrinol. Metab.* **86**, 5600–5608

20. Vottero, A., Kino, T., Combe, H., Lecomte, P. & Chrousos, G.P. (2002) A novel, C-terminal dominant negative mutation of the GR causes familial glucocorticoid resistance through abnormal interactions with p160 steroid receptor coactivators. *J. Clin. Endocrinol. Metab.* **87**, 2658–2667
21. Ruiz, M., Lind, U., Gafvels, M., Eggertsen, G., Carlstedt-Duke, J., Nilsson, L., Holtmann, M., Stierna, P., Wikstrom, A.C. & Werner, S. (2001) Characterization of two novel mutations in the glucocorticoid receptor gene in patients with primary cortisol resistance. *Clin. Endocrinol. (Oxford)* **55**, 363–371
22. Mendonca, B.B., Leite, M.V., de Castro, M., Kino, T., Elias, L.L., Bachega, T.A., Arnhold, I.J., Chrousos, G.P. & Latronico, A.C. (2002) Female pseudohermaphroditism caused by a novel homozygous missense mutation of the GR gene. *J. Clin. Endocrinol. Metab.* **87**, 1805–1809
23. Chang, S.S., Grunder, S., Hanukoglu, A., Rosler, A., Mathew, P.M., Hanukoglu, I., Schild, L., Lu, Y., Shimkets, R.A., Nelson-Williams, C. et al.(1996) Mutations in subunits of the epithelial sodium channel cause salt wasting with hyperkalaemic acidosis, pseudohypoaldosteronism type 1. *Nat. Genet.* **12**, 248–253
24. Strautnieks, S.S., Thompson, R.J., Gardiner, R.M. & Chung, E. (1996) A novel splice-site mutation in the γ subunit of the epithelial sodium channel gene in three pseudohypoaldosteronism type 1 families. *Nat. Genet.* **13**, 248–250
25. Cheek, D.B. & Perry, J.W. (1958) A salt wasting syndrome in infancy. *Arch. Dis. Child.* **22**, 252–256
26. Zennaro, M.C., Borensztein, P., Jeunemaitre, X., Armanini, D. & Soubrier, F. (1994) No alteration in the primary structure of the mineralocorticoid receptor in a family with pseudohypoaldosteronism. *J. Clin. Endocrinol. Metab.* **79**, 32–38
27. Arai, K., Tsigos, C., Suzuki, Y., Listwak, S., Zachman, K., Zangeneh, F., Rapaport, R., Chanoine, J.P. & Chrousos, G.P. (1995) No apparent mineralocorticoid receptor defect in a series of sporadic cases of pseudohypoaldosteronism. *J. Clin. Endocrinol. Metab.* **80**, 814–817
28. Arai, K., Zachman, K., Shibasaki, T. & Chrousos, G.P. (1999) Polymorphisms of amiloride-sensitive sodium channel subunits in five sporadic cases of pseudohypoaldosteronism: do they have pathologic potential? *J. Clin. Endocrinol. Metab.* **84**, 2434–2437
29. Tajima, T., Kitagawa, H., Yokoya, S., Tachibana, K., Adachi, M., Nakae, J., Suwa, S., Katoh, S. & Fujieda, K. (2000) A novel missense mutation of mineralocorticoid receptor gene in one Japanese family with a renal form of pseudohypoaldosteronism type 1. *J. Clin. Endocrinol. Metab.* **85**, 4690–4694
30. Viemann, M., Peter, M., Lopez-Siguero, J.P., Simic-Schleicher, G. & Sippell, W.G. (2001) Evidence for genetic heterogeneity of pseudohypoaldosteronism type 1: identification of a novel mutation in the human mineralocorticoid receptor in one sporadic case and no mutations in two autosomal dominant kindreds. *J. Clin. Endocrinol. Metab.* **86**, 2056–2059
31. Riepe, F.G., Krone, N., Morlot, M., Ludwig, M., Sippell, W.G. & Partsch, C.J. (2003) Identification of a novel mutation in the human mineralocorticoid receptor gene in a german family with autosomal-dominant pseudohypoaldosteronism type 1: further evidence for marked interindividual clinical heterogeneity. *J. Clin. Endocrinol. Metab.* **88**, 1683–1686
32. Sartorato, P., Lapeyraque, A.L., Armanini, D., Kuhnle, U., Khaldi, Y., Salomon, R., Abadie, V., Di Battista, E., Naselli, A., Racine, A. et al. (2003) Different inactivating mutations of the mineralocorticoid receptor in fourteen families affected by type I pseudohypoaldosteronism. *J. Clin. Endocrinol. Metab.* **88**, 2508–2517
33. Geller, D.S., Farhi, A., Pinkerton, N., Fradley, M., Moritz, M., Spitzer, A., Meinke, G., Tsai, F.T., Sigler, P.B. & Lifton, R.P. (2000) Activating mineralocorticoid receptor mutation in hypertension exacerbated by pregnancy, *Science* **289**, 119–123
34. Rafestin-Oblin, M.E., Souque, A., Bocchi, B., Pinon, G., Fagart, J. & Vandewalle, A. (2003) The severe form of hypertension caused by the activating S810L mutation in the mineralocorticoid receptor is cortisone related. *Endocrinology* **144**, 528–533

11

Nuclear receptors in disease: the oestrogen receptors

Maria Nilsson, Karin Dahlman-Wright and Jan-Åke Gustafsson[1]

Departments of Biosciences and Medical Nutrition, Karolinska Institutet, Novum, SE-141 57 Huddinge, Sweden

Abstract

For several decades, it has been known that oestrogens are essential for human health. The discovery that there are two oestrogen receptors (ERs), ERα and ERβ, has facilitated our understanding of how the hormone exerts its physiological effects. The ERs belong to the family of ligand-activated nuclear receptors, which act by modulating the expression of target genes. Studies of ER-knockout (ERKO) mice have been instrumental in defining the relevance of a given receptor subtype in a certain tissue. Phenotypes displayed by ERKO mice suggest diseases in which dysfunctional ERs might be involved in aetiology and pathology. Association between single-nucleotide polymorphisms (SNPs) in ER genes and disease have been demonstrated in several cases. Selective ER modulators (SERMs), which are selective with regard to their effects in a certain cell type, already exist. Since oestrogen has effects in many tissues, the goal with a SERM is to provide beneficial effects in one target tissue while avoiding side effects in others. Refined SERMs will, in the future, provide improved therapeutic strategies for existing and novel indications.

[1]*To whom correspondence should be addressed (jan-ake.gustafsson@mednut.ki.se).*

The oestrogen receptors (ERs)

For several decades, it has been known that the steroid hormone oestrogen is involved in disease. As early as 1896, Dr George Beatson conducted experiments in which removal of the ovaries from post-menopausal women with advanced breast cancer, led to shrinkage of the tumour and improved prognosis (reviewed in [1]). In 1916, ovariectomy of mice with a propensity to develop mammary carcinoma was found to reduce the incidence of tumours. In 1923, it was discovered that oestrogenic hormones are produced by the ovaries. In the late 1950s, the existence of a receptor molecule that could bind 17β-oestradiol was discovered by Jensen and Jacobson (reviewed in [1]); in 1986, the first ER was cloned [2,3]. This receptor was regarded as the only existing ER for 10 years, until members of our laboratory discovered a second ER [4]. The two receptors are today known as ERα and ERβ respectively, and belong to the family of ligand-activated nuclear receptors. These receptors exert their effects by modulating the expression of target genes. ERα and ERβ show a high degree of similarity when compared at the amino-acid level (Figure 1); however, it is clear that the receptors display differences with regard to ligand-binding and transcriptional activation. The receptors can be detected in a broad spectrum of tissues. In some organs, both receptor subtypes are expressed at similar levels, whereas in others, one or the other subtype predominates. ERα is, for example, the major receptor in the liver, while ERβ is the main receptor in the intestines. Both receptors can be found in heart and bone. Also, both receptor subtypes may be present in the same tissue, but in different cell types. For instance, ERα is found in the stromal cells and ERβ in the epithelial cells of the prostate. Mice lacking one or both of the receptor subtypes have been, and still are, instrumental in increasing our understanding of the role of oestrogen and its receptors in physiology and disease [5]. The ER-knockout (ERKO) mice display varying phenotypes in bone, brain, mammary gland, immune system, cardiovascular system and

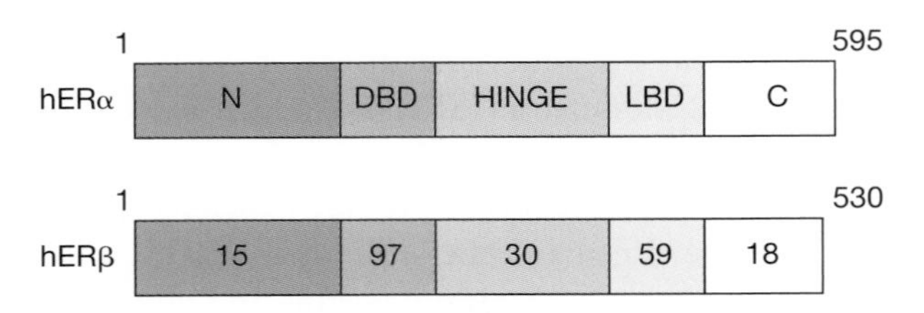

Figure 1. Comparison of the human ERs (hER) at the amino acid level
Percentage amino acid identity between the two receptor types is indicated in the model for hERβ. ERα is 595 amino acids long and ERβ WT (ERβ1) is 530 amino acids long. N, N-terminal domain; DBD, DNA-binding domain; HINGE, hinge domain, which bridges the DBD and ligand-binding domain (LBD); C, the very C-terminal domain, which harbours transactivation activity.

prostate, all of which will be discussed further in this review. Both male and female αERKO and αβERKO mice are completely infertile, whereas βERKO males exhibit normal fertility [5]; βERKO females show decreased fertility [6]. The infertility in the ERKO models can be partly accounted for by an inability to ovulate; in fact, both ERs are required for ovulation to occur efficiently.

ER modulators in physiology and disease

ER modulators, agonists and antagonists, have a widespread use in clinical practice today. The total world market for this class of drugs is worth billions of dollars. The introduction of ER antagonists for the treatment of hormone-dependent breast cancer represents a milestone in the treatment of this life-threatening disease. ER agonists are often used to alleviate the symptoms associated with post-menopausal syndrome. However, the risk–benefit profile of this substitution therapy needs to be considered.

The following sections focus on some selected diseases in which oestrogen and its receptors have been implicated.

ERs and breast cancer

It is generally believed that breast tumours, at least initially, are dependent on the stimulatory effects of oestrogens, directly or indirectly, and it is thought that oestrogens act by inducing the expression of paracrine growth factors and their receptors. However, many breast tumours eventually progress to an oestrogen-independent growth phenotype. Tamoxifen and similar anti-oestrogens are currently the first-line therapy for treatment of hormone-dependent breast cancer [1].

Various ER transcripts have been found in breast carcinomas [7,8] and protein products corresponding to variant ERs have been described [9]. Splicing of ERα precursor RNA frequently leads to variants, lacking one or more exons, that have been associated with breast cancer progression. The most frequent splice variants are exon 4 or exon 7 deletions [10]. Exon 4 comprises the hinge region and the very N-terminal part of the ligand-binding domain (LBD), and exon 7 contains part of the LBD. Of these two, only the exon 7 deletion has so far been detected at the protein level [9]. Single exon deletions in ERα can be found in a majority of normal breast tissues, while an increased frequency of multiple exon deletions is found in tumours [9]. Normal and cancerous tissues display a variety of distinct profiles regarding ERα wild-type (WT), ERβ WT, and ERα and ERβ splice variants at both mRNA and protein levels. This heterogeneity in ER isoform profiles is suggested to result in variations in oestrogen signalling, and might affect breast cancer risk, hormone-responsiveness and survival [9]. Some data suggest that the ERβ1 (ERβ WT) transcript is down-regulated in breast tumorigenesis [11–13], but this was not shown to be consistent for all tumour grades [14]. A higher level of ERβ2 (ERβ CX), compared with the ERβ1 transcript, was reported in human breast cancer cell lines, as well as in breast tumours and normal breast tissue [12–15]. ERβ CX is a variant of ERβ WT in which the last exon (LIV; see Figure 2B) is spliced away and replaced with the CX exon. We have shown that methylation of the ERβ promoter is inversely correlated with ERβ expression in breast cancer tumours and cell lines [15]. Since promoter methylation is frequently observed in cancer, these data suggest that ERβ is a possible tumour-suppressor gene.

ERβ is found in ductal, lobular epithelial and stromal cells of the rodent mammary gland (reviewed in [16]). ERα, on the other hand, is only found in the ductal and lobular epithelial cells, and not in the stroma. The αERKO mice display severe retardation of mammary gland development and show rudimentary glands at adulthood [16]. Studies of the βERKO mice indicate that ERβ is not necessary for ductal growth of the gland, but seems to be important for the organization and adhesion of epithelial cells [16].

ERs and prostate cancer

Prostate cancer is the most frequently diagnosed malignancy and the second most common cause of death among men in the U.S.A. Prostate cancer is age-dependent and the incidence increases after the age of 40. The growth and development of the prostate is under endocrine control and both androgens and oestrogens play important roles [17]. Androgens are essential for stimulating normal development, growth and secretory activities of the prostate, whereas oestrogens are generally regarded as inhibitors of growth. Combined androgen and oestrogen treatment has been shown to induce prostatic dysplasia and adenocarcinoma [18]. These results demonstrate that malignant changes in the prostate gland may be dependent upon both androgenic and oestrogenic effects. Oestrogens have been shown to induce growth of LNCaP prostate cancer cells, which can be inhibited by anti-oestrogens [19]. However, this cell line contains a mutant androgen receptor (AR), which results in promiscuity in hormone binding and receptor activation, making interpretation of these results difficult. In an androgen-decreased environment, oestrogen, through ERα, causes benign prostate hyperplasia, dysplasia and cancer [20]. Studies have shown that anti-oestrogens and specific ER modulators (SERMs) can delay and suppress prostate carcinogenesis [20]. Both ER subtypes are found in the ventral prostate, but are located in different cell types [16]. ERα is found in the stromal cells and ERβ in the epithelial cells, in which ERα is absent. The oestrogenic effects in the prostate may therefore be exerted by both ERs, but in different cells. Studies of transgenic mice show interesting phenotypes with hyperplastic prostates in both βERKO and aromatase-knockout (ARKO) mice, but not in αERKO mice [16]. In prostates from βERKO mice, most epithelial cells express the proliferation antigen Ki-67, and the tissue contains several hyperplastic foci [16]. In contrast, the prostate epithelium of their WT littermates show no hyperplasia and with only a few cells expressing Ki-67 [16]. The hyperplasia is thought to be directly caused by lack of ERβ and its anti-proliferative function in the prostate. The hyperplasia in ARKO mice may also be due to increased androgen and prolactin levels, secondary effects caused by the general loss of aromatase, the enzyme that converts androgens into oestrogens [16].

ERs and cardiovascular disease

Women present a higher risk for cardiovascular disease (CVD) after the onset of menopause, a phenomenon thought to be due to the loss of endogenous oestrogen. Accordingly, in some studies, reduced cardiovascular risks have been observed in subjects undergoing hormonal replacement therapy (HRT). Oestrogens, acting via ERs in the cardiovascular system [21], are thought to be important in prevention of CVD in women. Oestrogens have favourable effects on lipid profile, tone of vascular smooth muscle cells and fibrinogen levels [22]. When prescribed alone, however, oestrogen increases the risk of endometrial cancer and is therefore taken in combination with progestins, which are anti-proliferative in the uterus. Importantly, however, an oestrogen–progestin arm of the first prospective study of oestrogen and oestrogen–progestin for prevention of CVD was terminated early owing to an unacceptable risk profile. The Women's Health Initiative (WHI) reported that oestrogen in combination with progestin does not confer cardiac protection and may even increase the risk of CVD among healthy post-menopausal women, especially during the first year of treatment [23]. Furthermore, there was an increased risk of ischaemic stroke in generally healthy post-menopausal women [24]. However, the oestrogen-alone arm of the study is on-going. Results from the study of ERKO mice suggest that ERα is important in the pathophysiology of the vessel wall [25]. The βERKO mice display a phenotype with abnormalities in ion-channel function and an age-related sustained systolic and diastolic hypertension [25].

ERs and osteoporosis

Oestrogen and its receptors are known to be important in the regulation of bone metabolism. Oestrogen deficiency beginning at the menopause is a major pathogenic factor in the development of osteoporosis in post-menopausal women. The ERs are expressed in most cell types in bone [26]. A male patient with a non-functional ERα gene showed abnormal post-pubertal bone elongation [27]. Mice lacking the ERα gene show minor skeletal abnormalities with reduced longitudinal bone growth and small reductions in bone mineral density (BMD) [26]. Studies of female βERKO mice, which lack ERβ, indicate that ERβ is responsible for the repression of the growth-promoting effect of oestrogen on bone mediated via ERα [26].

ERs and diseases of the central nervous system (CNS)

ERs have also been implicated in various disorders of the brain. The receptors are expressed in the CNS, where they are thought to play important roles [16]. The distribution pattern suggests different functions for the two receptors. βERKO mice show an interesting phenotype with severe neuronal deficiency in the cortex, revealing an important role for ERβ in neuronal migration [16,28]. αERKO mice show no morphological abnormalities in the brain. Oestrogen has been proposed to act as a neuroprotectant. Deprivation of oestrogen as a result of the menopause is associated with an increased risk of Alzheimer's disease and Parkinson's disease [29]. Oestrogen replacement therapy may reduce this risk in both men and women.

ERs and diseases of the immune system

Oestrogen and its receptors also play important roles in the immune system. Ovariectomized mice show splenomegaly [30] and an increased production of haematopoietic cells [31]. Proliferation of pluripotent bone marrow stem cells is negatively regulated by oestrogen [31]. βERKO mice develop pronounced splenomegaly by 1.5 years of age [31], a phenomenon that is much more severe in females than in males. Interestingly, the absence of ERβ results in a myeloproliferative disease resembling human chronic myeloid leukaemia with lymphoid blast crisis. These intriguing results suggest a role for ERβ in regulating the pluripotent haematopoietic progenitor cells and make the βERKO mice a potential model for lymphoid and myeloid leukaemia [31].

ER polymorphisms and mutations in relation to disease

A few studies have been published in which the ERα and ERβ genes have been screened to identify mutations which change the amino acid sequence of the proteins, and are therefore candidates to change receptor function. Various ERα variants, including deletion variants, have been found in breast carcinomas [9]. A point mutation in the ERα gene that generates a stop codon, resulting in a truncated ERα protein, has been reported in one male patient [27]. This patient is suffering from osteoporosis and infertility. In a systematic mutation screening of ERβ in probands of different mass extremes, five different genetic variants were identified [32]. Our laboratory has screened approx. 50 patients diagnosed with various infertility syndromes such as polycystic ovary syndrome, premature ovarian failure and endometriosis, for mutations in the ERβ exons, only to reveal a few new variations, none of which cause a change in the primary structure of the protein (M. Nilsson and M. Zelada-Hedman unpublished work). We have also screened 34 primary breast cancer specimens without detecting any novel variants (M. Nilsson, unpublished work).

Single-nucleotide polymorphisms (SNPs) are base-pair changes that exist naturally in a population. Usually, variants that occur at an allele frequency exceeding 1% are referred to as SNPs. SNPs provide important tools for human genetic studies. A set of polymorphisms are scored in cases and controls. If an association is detected, i.e. a specific nucleotide is more common at one position in the gene in cases compared with controls, this indicates that the polymorphism itself, or some other change in the same or neighbouring genes, is related to the disease. A number of polymorphisms have been reported in the ERs. An overview of commonly scored ERα and ERβ variants is given in Figure 2. Generally, these SNPs do not change the amino acid sequence of the resulting ER proteins and thus are not likely to change receptor function. However, that they change the expression level of the receptor proteins cannot be excluded. A number of case-control studies have investigated a possible association between ERα or ERβ SNPs and disease. Studies of coronary artery disease (CAD) patients failed to show an association with the ERα variants B and C (see Figure 2A) [33]. Conflicting results have been reported for association between the ERα variants E and F (Figure 2A) and BMD [34], probably because of differences in screened populations. Another study reported that ERα variant B (Figure 2A) might be a risk factor for prostate cancer [35]. We have shown an association between ERβ SNPs I and J (Figure 2B) and bulimic patients [36], and others have shown an association between ERβ D and anorexia nervosa [37].

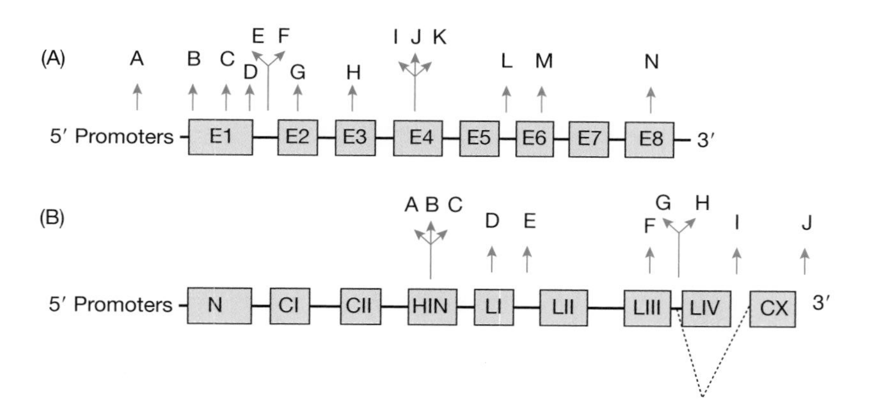

Figure 2. SNPs at the human ERα (A) and ERβ (B) gene loci

(**A**) E1–E8, exons 1–8; A, $(TA)_n$ repeats (17 alleles); B, codon 10 C/T (*Msp*I RFLP); C, codon 87 G/C (*Bst*UI RFLP); D, codon 100 C/G; E, intron 1 A/G (*Xba*I RFLP); F, intron 1 C/T (*Pvu*II RFLP); G, codon 160 G/T; H, codon 243 C/T; I, codon 309 C/T; J, codon 311 G/A; K, codon 325 C/G; L, $(CA)_n$ repeats (9 alleles); M, codon 425 C/T; N, codon 594 A/G. (**B**) N, N-terminal exon, CI, first part (finger) of the DNA-binding domain; CII, second part (finger) of the DNA-binding domain; HIN, hinge exon; LI–LIV, ligand-binding domain exons; CX, CX exon; A, 661 A/G; B, 809 (del21); C, 846 G/A; D, 1082 G/A; E, $(CA)_n$ repeat; F, 1421 T/C; G, ERβ LIV −68 C/T; H, ERβ LIV −4 A/G; I, 1730 G/A; J, ERβ CX +56 G/A. The ERβ LIV −68 and −4 polymorphisms are located 68 and 4 nucleotides 5′ of exon LIV respectively, and the ERβ CX +56 is located 56 nucleotides 3′ of exon CX.

Interestingly, ERβ is located on chromosome 14, in a recently identified region that meets the criteria for genome-wide suggestive linkage with bulimia nervosa [38]. However, a German study did not provide evidence for an association between bulimia nervosa and ERβ polymorphisms D and I [32]. Other studies have investigated the possible association between a dinucleotide repeat polymorphism (E in Figure 2B) located in the flanking region of the human ERβ gene and various clinical parameters. A possible association between this polymorphism and hypertension in Japanese women was reported in [39]. The dinucleotide repeat was also suggested to be associated with BMD [40] and androgen levels in women [41], but not with autoimmune thyroid diseases in another study [42]. Association studies of the same ERβ gene polymorphism found a correlation with higher BMD in pre-menopausal, but not post-menopausal Chinese women [43], suggesting that the ERβ gene may have a modulatory role in bone metabolism in young adulthood. Six different polymorphisms in ERβ (B, C, D, E, H and I) were studied in a sporadic breast cancer material; however, no differences were found in the allelic distribution of the six studied polymorphisms between the breast cancer and control groups [44].

Conclusion

The ERs are known to have important roles in many different diseases, including osteoporosis, prostate cancer, breast cancer, cardiovascular disease and diseases of the CNS. Recently, other interesting functions of the ERs have been proposed. Studies of mice revealed that absence of ERβ results in a myeloproliferative disease resembling human chronic myeloid leukaemia with lymphoid blast crisis. This might indicate a role for ERβ in regulating the differentiation of pluripotent haematopoietic progenitor cells. Future association studies on large populations will contribute to our understanding of the role of ERα and ERβ for the development of disease in humans. We are expecting an explosion in our understanding of the molecular effects of oestrogen in target tissues in relation to physiology and disease. This will be possible using modern gene expression profiling and proteomics technologies. Furthermore, receptor-selective ligands, which are currently being developed, will provide important tools to understand ER biology and studies in ERα- and/or ERβ-deficient mice will clarify the importance of a specific receptor for a given effect. These studies will aid the development of SERMs with well-characterized and optimal effects in oestrogen-responsive tissues. SERMs are already available; however, existing SERMs are modest in their desired tissue-selective activity and have undesired side effects. Therefore a great need exists for novel SERMs with a more refined therapeutic profile with optimal selectivity and activity in oestrogen-responsive tissues.

Summary

- *The ERs have been found to play important roles in many different diseases, including osteoporosis, prostate cancer, breast cancer and CVD. Oestrogen is suggested to have a neuroprotective role. The onset of the menopause most probably leads to an increased risk of neurodegenerative disorders, such as Parkinson's and Alzheimer's diseases, in women.*
- *The ERKO mice display interesting phenotypes, including alterations in bone, breast, CNS, immune system, reproductive organs and prostate.*
- *ER variants have been identified, only a few of which result in changes in amino acid sequence.*
- *Association studies have shown correlations between ERβ polymorphisms and bulimia, anorexia, BMD and hypertension.*
- *The development of SERMs with well-characterized and optimal effects in oestrogen-responsive tissues will be very important in the treatment of ER-dependent disorders.*

References

1. Jensen, E.V. & Jordan, V.C. (2003) The estrogen receptor: a model for molecular medicine. *Clin. Cancer Res.* **9**, 1980–1989
2. Green, S., Walter, P., Kumar, V., Krust, A., Bornert, J.M., Argos, P. & Chambon, P. (1986) Human oestrogen receptor cDNA: sequence, expression and homology to v-erb-A. *Nature (London)* **320**, 134–139
3. Greene, G.L., Gilna, P., Waterfield, M., Baker, A., Hort, Y. & Shine, J. (1986) Sequence and expression of human estrogen receptor complementary DNA. *Science* **231**, 1150–1154
4. Kuiper, G.G., Enmark, E., Pelto-Huikko, M., Nilsson, S. & Gustafsson, J.-Å. (1996) Cloning of a novel receptor expressed in rat prostate and ovary. *Proc. Natl. Acad. Sci. U.S.A.* **93**, 5925–5930
5. Couse, J.F. & Korach, K.S. (1999) Estrogen receptor null mice: what have we learned and where will they lead us? *Endocr. Rev.* **20**, 358–417
6. Dupont, S., Krust, A., Gansmuller, A., Dierich, A., Chambon, P. & Mark, M. (2000) Effect of single and compound knockouts of estrogen receptors α (ERα) and β (ERβ) on mouse reproductive phenotypes. *Development* **127**, 4277–4291
7. Pfeffer, U., Fecarotta, E. & Vidali, G. (1995) Coexpression of multiple estrogen receptor variant messenger RNAs in normal and neoplastic breast tissues and in MCF-7 cells. *Cancer Res.* **55**, 2158–2165
8. Gotteland, M., Desauty, G., Delarue, J.C., Liu, L. & May, E. (1995) Human estrogen receptor messenger RNA variants in both normal and tumor breast tissues. *Mol. Cell. Endocrinol.* **112**, 1–13
9. Poola, I. & Speirs, V. (2001) Expression of alternatively spliced estrogen receptor a mRNAs is increased in breast cancer tissues. *J. Steroid Biochem. Mol. Biol.* **78**, 459–469
10. Ferro, P., Forlani, A., Muselli, M. & Pfeffer, U. (2003) Alternative splicing of the human estrogen receptor a primary transcript: mechanisms of exon skipping. *Int. J. Mol. Med.* **12**, 355–363
11. Leygue, E., Dotzlaw, H., Watson, P.H. & Murphy, L.C. (1998) Altered estrogen receptor α and β messenger RNA expression during human breast tumorigenesis. *Cancer Res.* **58**, 3197–3201
12. Leygue, E., Dotzlaw, H., Watson, P.H. & Murphy, L.C. (1999) Expression of estrogen receptor β1, β2, and β5 messenger RNAs in human breast tissue. *Cancer Res.* **59**, 1175–1179

13. Iwao, K., Miyoshi, Y., Egawa, C., Ikeda, N. & Noguchi, S. (2000) Quantitative analysis of estrogen receptor-β mRNA and its variants in human breast cancers. *Int. J. Cancer* **88**, 733–736
14. Palmieri, C., Cheng, G.J., Saji, S., Zelada-Hedman, M., Warri, A., Weihua, Z., Van Noorden, S., Wahlstrom, T., Coombes, R.C., Warner, M. & Gustafsson, J.Å. (2002) Estrogen receptor β in breast cancer. *Endocr. Relat. Cancer* **9**, 1–13
15. Zhao, C., Lam, E.W., Sunters, A., Enmark, E., De Bella, M.T., Coombes, R.C., Gustafsson, J.-Å. & Dahlman-Wright, K. (2003) Expression of estrogen receptor β isoforms in normal breast epithelial cells and breast cancer: regulation by methylation. *Oncogene* **22**, 7600–7606
16. Weihua, Z., Andersson, S., Cheng, G., Simpson, E.R., Warner, M. & Gustafsson, J.-Å. (2003) Update on estrogen signaling. *FEBS Lett.* **546**, 17–24
17. Hsing, A.W., Reichardt, J.K. & Stanczyk, F.Z. (2002) Hormones and prostate cancer: current perspectives and future directions. *Prostate* **52**, 213–235
18. Risbridger, G.P., Bianco, J.J., Ellem, S.J. & McPherson, S.J. (2003) Oestrogens and prostate cancer. *Endocr. Relat. Cancer* **10**, 187–191
19. Castagnetta, L.A., Miceli, M.D., Sorci, C.M., Pfeffer, U., Farruggio, R., Oliveri, G., Calabro, M. & Carruba, G. (1995) Growth of LNCaP human prostate cancer cells is stimulated by estradiol via its own receptor. *Endocrinology* **136**, 2309–2319
20. Steiner, M.S. & Raghow, S. (2003) Antiestrogens and selective estrogen receptor modulators reduce prostate cancer risk. *World J. Urol.* **21**, 31–36
21. Mendelsohn, M.E. & Karas, R.H. (1999) The protective effects of estrogen on the cardiovascular system. *N. Engl. J. Med.* **340**, 1801–1811
22. Khan, N.S. & Malhotra, S. (2003) Effect of hormone replacement therapy on cardiovascular disease: current opinion. *Expert Opin. Pharmacother.* **4**, 667–674
23. Hays, J., Ockene, J.K., Brunner, R.L., Kotchen, J.M., Manson, J.E., Patterson, R.E., Aragaki, A.K., Shumaker, S.A., Brzyski, R.G., LaCroix, A.Z. et al. (2003) Effects of estrogen plus progestin on health-related quality of life. *N. Engl. J. Med.* **348**, 1839–1854
24. Manson, J.E., Hsia, J., Johnson, K.C., Rossouw, J.E., Assaf, A.R., Lasser, N.L., Trevisan, M., Black, H.R., Heckbert, S.R., Detrano, R. et al. (2003) Estrogen plus progestin and the risk of coronary heart disease. *N. Engl. J. Med.* **349**, 523–534
25. Otsuki, M., Gao, H., Dahlman-Wright, K., Ohlsson, C., Eguchi, N., Urade, Y. & Gustafsson, J.-Å. (2003) Specific regulation of lipocalin-type prostaglandin D synthase in mouse heart by estrogen receptor β. *Mol. Endocrinol.* **17**, 1844–1855
26. Sims, N.A., Dupont, S., Krust, A., Clement-Lacroix, P., Minet, D., Resche-Rigon, M., Gaillard-Kelly, M. & Baron, R. (2002) Deletion of estrogen receptors reveals a regulatory role for estrogen receptors-β in bone remodeling in females but not in males. *Bone* **30**, 18–25
27. Smith, E.P., Boyd, J., Frank, G.R., Takahashi, H., Cohen, R.M., Specker, B., Williams, T.C., Lubahn, D.B. & Korach, K.S. (1994) Estrogen resistance caused by a mutation in the estrogen-receptor gene in a man. *N. Engl. J. Med.* **331**, 1056–1061
28. Wang, L., Andersson, S., Warner, M. & Gustafsson, J.-Å. (2003) Estrogen receptor (ER)β knock-out mice reveal a role for ERβ in migration of cortical neurons in the developing brain. *Proc. Natl. Acad. Sci. U.S.A.* **100**, 703–708
29. Bhavnani, B.R. (2003) Estrogens and menopause: pharmacology of conjugated equine estrogens and their potential role in the prevention of neurodegenerative diseases such as Alzheimer's. *J. Steroid Biochem. Mol. Biol.* **85**, 473–482
30. Zhang, J., Pugh, T.D., Stebler, B., Ershler, W.B. & Keller, E.T. (1998) Orchiectomy increases bone marrow interleukin-6 levels in mice. *Calcif. Tissue Int.* **62**, 219–226
31. Shim, G.J., Wang, L., Andersson, S., Nagy, N., Kis, L.L., Zhang, Q., Makela, S., Warner, M. & Gustafsson, J.-Å. (2003) Disruption of the estrogen receptor β gene in mice causes myeloprolif-erative disease resembling chronic myeloid leukemia with lymphoid blast crisis. *Proc. Natl. Acad. Sci. U.S.A.* **100**, 6694–6699
32. Rosenkranz, K., Hinney, A., Ziegler, A., Hermann, H., Fichter, M., Mayer, H., Siegfried, W., Young, J.K., Remschmidt, H. & Hebebrand, J. (1998) Systematic mutation screening of the estrogen

receptor β gene in probands of different weight extremes: identification of several genetic variants. *J. Clin. Endocrinol. Metab.* **83**, 4524–4527

33. Evangelopoulos, D., Alevizaki, M., Lekakis, J., Cimponeriu, A., Papamichael, C., Kominakis, A., Kalofoutis, A. & Moutsatsou, P. (2003) Molecular analysis of the estrogen receptor α gene in men with coronary artery disease: association with disease status. *Clin. Chim. Acta* **331**, 37–44
34. Gennari, L., Becherini, L., Falchetti, A., Masi, L., Massart, F. & Brandi, M.L. (2002) Genetics of osteoporosis: role of steroid hormone receptor gene polymorphisms. *J. Steroid Biochem. Mol. Biol.* **81**, 1–24
35. Tanaka, Y., Sasaki, M., Kaneuchi, M., Shiina, H., Igawa, M. & Dahiya, R. (2003) Polymorphisms of estrogen receptor α in prostate cancer. *Mol. Carcinog.* **37**, 202–208
36. Nilsson, M., Naessen, S., Dahlman, I., Linden Hirschberg, A., Gustafsson, J.-Å. & Dahlman-Wright, K. (2004) Association of estrogen receptor β gene polymorphisms with bulimic disease in women. *Mol. Psychiatry* **9**, 28–34
37. Eastwood, H., Brown, K.M., Markovic, D. & Pieri, L.F. (2002) Variation in the ESR1 and ESR2 genes and genetic susceptibility to anorexia nervosa. *Mol. Psychiatry* **7**, 86–89
38. Bulik, C.M., Devlin, B., Bacanu, S.A., Thornton, L., Klump, K.L., Fichter, M.M., Halmi, K.A., Kaplan, A.S., Strober, M., Woodside, D.B. et al. (2003) Significant linkage on chromosome 10p in families with bulimia nervosa. *Am. J. Hum. Genet.* **72**, 200–207
39. Ogawa, S., Emi, M., Shiraki, M., Hosoi, T., Ouchi, Y. & Inoue, S. (2000) Association of estrogen receptor β (ESR2) gene polymorphism with blood pressure. *J. Hum. Genet.* **45**, 327–330
40. Ogawa, S., Hosoi, T., Shiraki, M., Orimo, H., Emi, M., Muramatsu, M., Ouchi, Y. & Inoue, S. (2000) Association of estrogen receptor β gene polymorphism with bone mineral density. *Biochem. Biophys. Res. Commun.* **269**, 537–541
41. Westberg, L., Baghaei, F., Rosmond, R., Hellstrand, M., Landen, M., Jansson, M., Holm, G., Bjorntorp, P. & Eriksson, E. (2001) Polymorphisms of the androgen receptor gene and the estrogen receptor β gene are associated with androgen levels in women. *J. Clin. Endocrinol. Metab.* **86**, 2562–2568
42. Ban, Y., Tozaki, T., Taniyama, M. & Tomita, M. (2001) Lack of association between estrogen receptor β dinucleotide repeat polymorphism and autoimmune thyroid diseases in Japanese patients. *BMC Med. Genet.* **2**, 1
43. Lau, H.H., Ho, A.Y., Luk, K.D. & Kung, A.W. (2002) Estrogen receptor β gene polymorphisms are associated with higher bone mineral density in premenopausal, but not postmenopausal southern Chinese women. *Bone* **31**, 276–281
44. Forsti, A., Zhao, C., Israelsson, E., Dahlman-Wright, K., Gustafsson, J.-Å. & Hemminki, K. (2003) Polymorphisms in the estrogen receptor β gene and risk of breast cancer: no association. *Breast Cancer Res. Treat.* **79**, 409–413

12

Nuclear receptors and human disease: thyroid receptor β, peroxisome-proliferator-activated receptor γ and orphan receptors

Mark Gurnell and V. Krishna K. Chatterjee[1]

Department of Medicine, University of Cambridge, Addenbrooke's Hospital, Hills Road, Cambridge CB2 2QQ, U.K.

Abstract

The nuclear receptor superfamily comprises a group of proteins that includes the molecular targets for classical steroid hormones such as glucocorticoids, androgens and vitamin D, together with a number of so-called 'orphan' receptors whose ligands and/or function remain to be determined. Many of the world's most commonly prescribed drugs act via nuclear receptors, attesting to their importance as therapeutic targets in human disease [for example, the novel anti-diabetic thiazolidinediones rosiglitazone and pioglitazone are high-affinity ligands for peroxisome-proliferator-activated receptor γ (PPARγ)]. The study of transgenic mice harbouring global and tissue-specific alterations in nuclear receptor genes has greatly enhanced our understanding of the roles that these receptors play in mammalian physiology. In many cases, these findings have been complemented by the study of human subjects harbouring naturally occurring mutations within the corresponding receptor, whereas in others, such studies have served to highlight important differences that exist

[1]*To whom correspondence should be addressed (e-mail kkc1@mole.bio.cam.ac.uk).*

between human and mouse physiology especially, for example, in relation to aspects of metabolism. Here we review the diverse clinical phenotypes that have been reported in subjects found to have germline mutations in thyroid hormone receptor β, PPARγ, hepatocyte nuclear factor 4α, small heterodimer partner, steroidogenic factor 1, DAX1, photoreceptor-specific nuclear receptor and NUR-related factor 1, and consider the molecular mechanisms through which aberrant signalling by mutant receptors might contribute to the pathogenesis of the associated disorders.

Introduction

Classical steroid hormones (e.g. glucocorticoids, mineralocorticoids, vitamin D) act principally within the cell nucleus to modulate target gene transcription through binding to specific receptors (e.g. glucocorticoid receptor, mineralocorticoid receptor, vitamin D receptor). These receptors are members of a broader nuclear receptor superfamily, which also includes proteins that are the targets for structurally unrelated ligands (e.g. thyroid hormone and retinoic acid), together with a large number of novel proteins, which were originally designated as 'orphan' receptors pending identification of their cognate ligand(s) [1]. However, the orphan status of several of these receptors has been challenged as evidence has begun to emerge of endogenous ligands that are capable of regulating receptor function at concentrations that approximate those found *in vivo*. Many of these molecules have also proved to be structurally distinct from steroid hormones, and bind with much lower affinities to receptors that are often more permissive. For example, a variety of fatty acid derivatives and eicosanoids have been shown to regulate peroxisome-proliferator-activated receptor γ (PPARγ) function.

As ligand-inducible transcription factors, nuclear receptors are organized in functional domains which are highly conserved among family members (Figure 1A): the N-terminal region often encodes an intrinsic transcriptional activation function (AF-1), a central DNA-binding domain (DBD) mediates receptor interaction with regulatory DNA sequences or response elements in target gene promoters, and the C-terminal region harbours the ligand-binding domain (LBD) and encompasses a powerful ligand-dependent transactivation function (AF-2). Inspection of the primary amino acid sequence encoding the central domain of virtually all nuclear receptors reveals the presence of two cysteine-rich motifs, each of which co-ordinates a zinc ion to form a 'finger-like' structure capable of directing sequence-specific DNA-binding (Figure 1A). The specificity of this interaction is mediated at least in part through the P-box, which lies at the base of the first zinc finger, with different amino acids within this region dictating DNA-response-element recognition. Residues within the P-box are encompassed within an α-helix that interacts directly with the major groove of DNA (see Chapter 5 in this volume). For some receptors, residues within an A-box form an additional α-helix which interacts with the minor groove of DNA (Figure 1A). Several receptors [e.g. glucocorticoid receptor, mineralocorticoid receptor, androgen receptor and hepatocyte nuclear factor 4α (HNF4α)] form homodimeric complexes on response elements consisting of palindromic arrangements of two hexanucleotide motifs, whereas others [e.g. thyroid hormone receptor (TR), retinoic acid receptor, vitamin D receptor and PPAR] interact with a tandem repeat arrangement of hexanucleotide motifs as a heterodimer with the retinoid X receptor, another member of the nuclear receptor family [2]. A third type of receptor–DNA interaction is exhibited by some orphan receptors [e.g. nerve growth factor inducible factor I-B and steroidogenic factor 1 (SF1)], which bind monomerically to extended response elements that include additional nucleotides 5′ to a core hexanucleotide motif.

The hallmark of nuclear receptors is their ability to modulate transcription in response to ligand occupancy. Many undergo a conformational change upon hormone binding which facilitates the recruitment of a complex of 'co-activators' [e.g. steroid receptor co-activator-1 (SRC-1), cAMP-response-element-binding protein (CREB)-binding protein (CBP) and p300/CBP-associating factor (pCAF)], which modify the chromatin structure so as to permit transcriptional activation (Figure 1B; see Chapter 6 in this volume). In the absence of ligand, a subset of receptors, which includes TR and retinoic acid receptor, is capable of mediating transcriptional repression through recruitment of a distinct co-repressor complex [e.g. silencing mediator for retinoid and thyroid hormone receptors (SMRT)/nuclear receptor co-repressor (NCoR), Sin3a and histone deacetylase; see Chapter 7 in this volume].

Mutations in nuclear receptor genes form the basis of a number of inherited human diseases (Table 1). In the majority of cases the link between nuclear receptor defects and a particular human phenotype has been identified using one of two lines of investigation: first, the 'candidate gene' approach with direct sequence analysis of a gene of interest in affected subjects (as exemplified by the identification of mutations in human PPARγ in subjects with severe insulin resistance and partial lipodystrophy); secondly, the 'reverse genetic' approach in which linkage studies have associated a disease with a chromosomal locus, following which positional cloning has identified a gene encoding a nuclear receptor [e.g. mutations in HNF4α in maturity onset diabetes of the young type 1 (MODY1)].

Human nuclear receptor gene defects may be broadly categorized into germline (inherited or sporadic) or somatic. The latter class includes mosaic expression of a mutant receptor and gene rearrangements occurring in tumour cells, generating, for example, the promyelocytic leukaemia–retinoic acid receptor α oncoprotein in acute promyelocytic leukaemia (APML) [3], or the recently described PAX8–PPARγ1 fusion protein in thyroid follicular neoplasia [4]. For simplicity, this chapter will focus on germline mutations in a selected group of nuclear receptors [TRβ, PPARγ, HNF4α, small heterodimer partner (SHP), SF1, dosage-sensitive sex reversal-adrenal hypoplasia congenita (AHC) critical region on the X chromosome gene 1 (DAX1), photoreceptor-specific nuclear

(A)

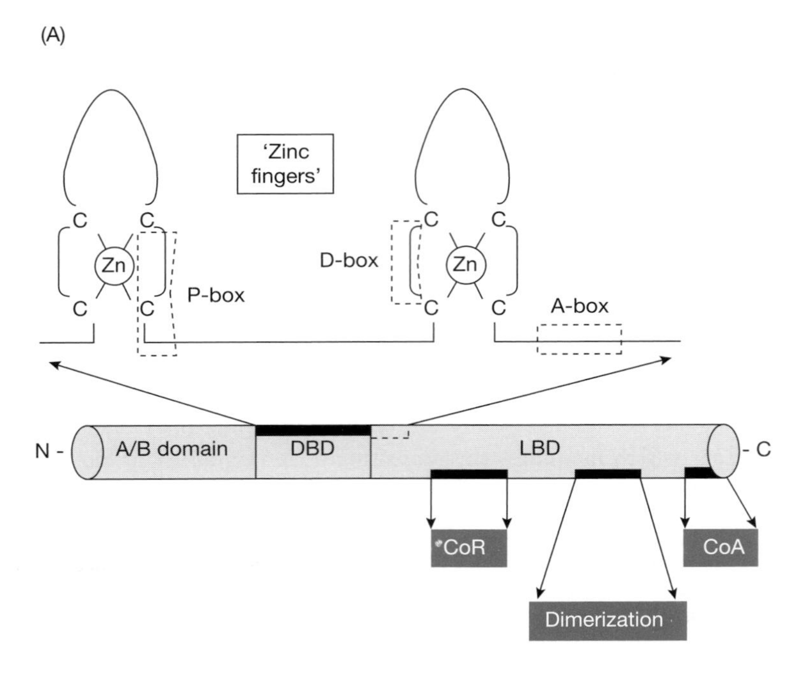
'Zinc fingers'
C
C
Zn
C
C
P-box
D-box
C
C
Zn
C
C
A-box
N -
A/B domain
DBD
LBD
- C
CoR
CoA
Dimerization

(B)

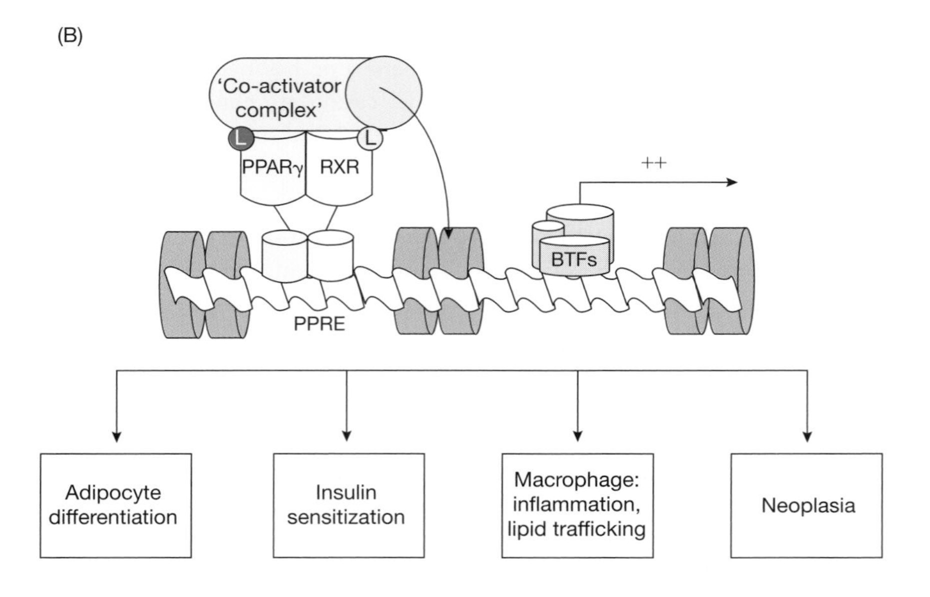
'Co-activator complex'
L
L
PPARγ
RXR
++
BTFs
PPRE
Adipocyte differentiation
Insulin sensitization
Macrophage: inflammation, lipid trafficking
Neoplasia

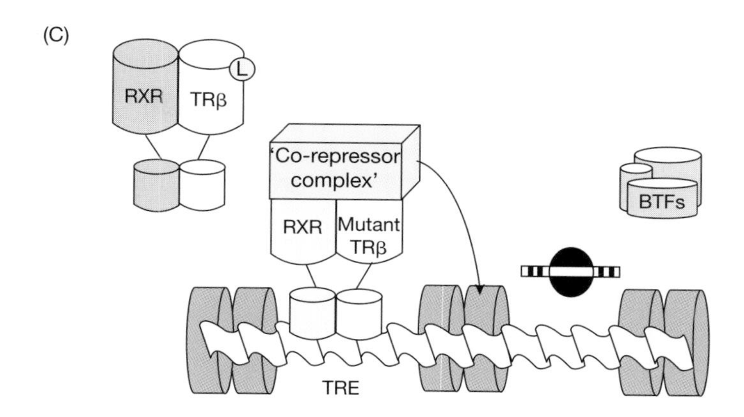

Figure 1. Nuclear receptor domains and structure–function relationships
(**A**) Schematic representation of the modular domain structure of the nuclear receptor superfamily. DNA-binding (DBD) and ligand-binding (LBD) domains, together with regions implicated in interaction with co-repressors (CoR) and co-activators (CoA), are indicated. The N-terminal A/B domain contains a constitutive (AF-1) activation function, whereas the LBD encompasses ligand-dependent transactivation (AF-2). Two 'zinc-finger' motifs interact directly with target gene-response elements, with residues contained within the P- and A-boxes, facilitating interaction with the major and minor groove of DNA respectively. (**B**) Addition of ligand (L) to the DNA-bound PPARγ–retinoid X receptor (RXR) heterodimer promotes recruitment of a transcriptional 'co-activator' protein complex, which in turn modulates the transcription of target genes regulating different biological processes. (**C**) Dominant-negative human TRβ mutants retain the ability to heterodimerize with retinoid X receptor on DNA, but fail to bind ligand and/or recruit transcriptional co-activators. The heterodimer recruits a 'co-repressor' protein complex, leading to silencing of target gene transcription. PPRE, PPAR-response element; TRE, thyroid-response element; BTFs, basal transcription factors.

receptor (PNR) and NUR-related factor 1 (NURR1)], highlighting the advances in our understanding of nuclear receptor biology that have been made through the study of these comparatively rare human disorders.

TRβ and the syndrome of resistance to thyroid hormone (RTH)

Background

In mammals, the TR exists in two major isoforms, TRα (NR1A1) and TRβ (NR1A2), which are encoded by genes on chromosomes 17 and 3 respectively. Both gene loci give rise to two protein products each, TRα1 and TRα2 (which differ at the C-terminal end, with only the former able to bind ligand) and TRβ1 and TRβ2 (which differ at the N-terminal end, such that both retain the ability to bind T3 with high affinity). These isoforms exhibit differing tissue expression patterns: for example, TRα1 is the main species in the myocardium and skeletal muscle; TRα2, which may act as a functional antagonist of other TR isoforms, is expressed in a variety of tissues (e.g. brain and testis); TRβ1 predominates in the liver and kidney; both TRβ1 and TRβ2 are highly

Table 1. Disorders associated with mutations in human nuclear receptors

VDR, vitamin D receptor; ER, oestrogen receptor; GR, glucocorticoid receptor; MR, mineralocorticoid receptor; AR, androgen receptor; ADHD, attention-deficit hyperactivity disorder; SHP, small heterodimer partner; PNR, photoreceptor-specific nuclear receptor; NURR1, NUR-related factor 1. Modes of inheritance: AD, autosomal dominant; AR, autosomal recessive; S, sporadic; XL, X-linked.

Receptor	Disorder	Clinical features	Mode of inheritance
TRβ	Resistance to thyroid hormone	Goitre, tachycardia, failure to thrive, ADHD	AD, AR, S
VDR	Hereditary vitamin D-dependent rickets (type II)	Hypocalcaemia, rickets, alopecia	AR
ERα	Oestrogen resistance	Tall stature, delayed epiphyseal fusion, osteoporosis (male)	AR
GR	Glucocorticoid resistance	Fatigue, hypertension, hyperandrogenism, infertility	AD, AR, S
MR	Pseudohypoaldosteronism	Hypotension, salt loss	AD
AR	Androgen-insensitivity syndrome	Partial or complete failure of masculinization	XL
PPARγ*	PPARγ ligand resistance	Partial lipodystrophy, insulin resistance, hypertension	AD
HNF4α	Maturity onset diabetes of the young type 1	Early-onset type 2 diabetes	AD
SHP		Mild obesity	AD
SF1		Primary adrenal failure, XY sex reversal	AD, AR
DAX1	Adrenal hypoplasia congenita	Primary adrenal failure, hypogonadotrophic hypogonadism, impaired spermatogenesis	XL
PNR	Enhanced S cone syndrome	Increased sensitivity to blue light, visual loss	AR
NURR1	(i) Familial Parkinson's syndrome	Resting tremor, bradykinesia, cogwheel rigidity	AD
	(ii) Schizophrenia/manic-depressive disorder		AD

*A second variant P115Q has been reported in association with obesity in four subjects.

expressed in the hypothalamus and pituitary. Although studies in mice have greatly facilitated our understanding of the differing roles of these receptor subtypes in normal physiology and have therefore provided clues to the human phenotypes that one might anticipate in subjects harbouring mutant TRs, only one disorder consequent on aberrant TR signalling has been identified hitherto in humans, the syndrome of RTH. This condition, first described more than three decades ago by Refetoff and colleagues [5], is characterized by elevated circulating free thyroid hormone levels with a failure to suppress pituitary thyroid-stimulating hormone secretion and variable refractoriness to hormone action in the periphery [6]. In the majority of cases, affected individuals are heterozygous for diverse mutations in the LBD of TRβ (occurring within a region common to both TRβ1 and TRβ2 variants).

Molecular genetics and mutant receptor properties in RTH

Shortly after the cloning of the TRα and TRβ genes, Usala and colleagues [7] reported tight linkage between RTH and the TRβ gene locus in a kindred with generalized resistance to thyroid hormone (GRTH; see below). Subsequently, >100 different mutations have been described in more than 500 RTH kindreds worldwide, with the majority being inherited in an autosomal dominant manner, but with a significant minority (20–25%) arising sporadically. However, a small number of subjects exist in whom no mutation in TRβ can be demonstrated despite compelling clinical and biochemical evidence for the disorder (so-called 'TRβ-negative RTH'), suggesting that abnormalities in other genes (e.g. transcriptional cofactors) can mimic the RTH phenotype. Data from animal studies support this notion, with mice harbouring deletion of SRC-1 exhibiting an RTH phenotype [8], as do animals doubly heterozygous for disruptions of SRC-1 and the homologous nuclear receptor co-activator transcriptional intermediary factor-2 (TIF-2) [9]. On the other hand, mice harbouring a mutation in TRα1 exhibit a phenotype dissimilar to RTH.

To date, virtually all of the mutations reported in RTH cluster within three regions of the LBD, often referred to as the 'codon 200' (residues 234–282), 'codon 300' (residues 310–353) and 'codon 400' (residues 429–461) mutations. In keeping with their location, mutant TRβ exhibit impaired hormone-binding and transcriptional activity. In addition, the mutant receptors are capable of inhibiting the action of their wild-type (WT) counterpart when co-expressed (Figure 1C) [10]. Clinical observations from two unusual cases of RTH provide evidence in support of the 'dominant-negative' inhibitory effects of mutant TRβ *in vivo*: in a unique family with recessively inherited RTH, individuals who were heterozygous for a deletion of one allele of the TRβ gene were clinically and biochemically unaffected [5], an observation which is consistent with findings in heterozygous TRβ-knockout mice [11]; conversely, a child exhibiting homozygosity for a dominant-negative TRβ mutant demonstrated a severe clinical phenotype with biochemical evidence of extreme resistance to thyroid hormone action [12]. Indeed, this critical requirement for dominant-negative activity by mutant TRβ in the pathogenesis of the disorder may explain why other regions of the receptor appear to be devoid of naturally occurring mutations (so-called 'cold areas'), reflecting the fact that a TRβ mutant must retain certain key properties (e.g. ability to heterodimerize with retinoid X receptor, bind to DNA and recruit transcriptional co-repressor) in order to induce the phenotype of RTH (Figure 1C).

Clinical features and management in RTH

A palpable goitre is the commonest presenting feature in RTH, and often triggers thyroid function tests, which may then reveal the distinctive biochemical signature of the disorder. Many of these individuals are otherwise asymptomatic and are deemed to have 'generalized' resistance (GRTH), a 'euthyroid' state in which high thyroid hormone levels are thought to compensate for global tissue resistance. However, the clinical manifestations of RTH are highly variable (Table 2), and other subjects may exhibit frank thyrotoxic features (e.g. failure to thrive in childhood, low body-mass index, tachycardia or dysrhythmia), suggesting predominant pituitary resistance to thyroid hormone (PRTH), a state in which peripheral tissues retaining 'normal' sensitivity to T_3 are exposed to high circulating hormone levels. Both GRTH and PRTH are associated with TRβ mutations, indicating that the two disorders represent phenotypic variants of a single genetic entity [13]. Moreover, whereas the clinical distinction between GRTH and PRTH remains useful in guiding management, there is significant overlap between these 'states'. For example, tachycardia, hyperkinetic behaviour and anxiety have been documented in individuals deemed to have GRTH; conversely, serum-sex-hormone-binding globulin (a hepatic index of thyroid hormone action) is typically normal in PRTH, suggesting that resistance is not solely confined to the hypothalamic/pituitary/thyroid axis (HPG axis).

In general, treatment of RTH is governed by the prevailing symptoms. Thus subjects with 'compensated' GRTH rarely require intervention. Surgery or radio-iodine to treat the biochemical disturbance or goitre frequently fails, and may serve only to complicate the clinical picture by rendering the RTH patient hypothyroid [14]. In contrast, subjects with overt thyrotoxic features

Table 2. Features of the syndrome of RTH

Elevated serum-free thyroid hormones
Unsuppressed TSH with enhanced bioactivity
Goitre
Tachycardia, atrial fibrillation, heart failure
Low body-mass index in childhood
Growth retardation, short stature
Attention-deficit hyperactivity disorder, low IQ
Ear, nose and throat infections, hearing loss
Osteopenia

(PRTH) may benefit from attempts to reduce pituitary thyroid-stimulating hormone output; most easily achieved by the administration of a thyroid hormone analogue such as TRIAC (3,3,5-tri-iodothyroacetic acid), which preferentially binds TRβ and exerts a predominantly pituitary thyromimetic effect, thereby lowering circulating free thyroid hormone levels. However, the picture is complex and what is beneficial for one tissue (e.g. heart) may be deleterious to another. A relatively hypothyroid state within the liver may cause hypercholesterolaemia. Moreover, temporal variations in receptor sensitivity have been observed, and accordingly periodic review of the treatment strategy in any given individual is mandatory.

PPARγ and the syndrome of PPARγ ligand resistance (PLR)

Background

PPARγ (NR1C3) was first characterized as a transcription factor that regulates target gene expression in adipocytes and induces pre-adipocyte differentiation [15]. The PPARγ gene (on chromosome 3) undergoes alternative splicing to generate two protein products: a long PPARγ2 isoform (containing 28 additional residues in the N-terminal domain) with a restricted pattern of expression (high levels in adipose tissue), and a shorter PPARγ1 variant, which is more widely distributed. Although it is little more than a decade since the receptor was first cloned, there is already an extensive body of data implicating it in a diverse array of biological processes that extend beyond the adipocyte. For example, PPARγ mediates inhibition of inflammatory cytokine production (interleukin-6 and tumour necrosis factor-α) from monocytes [16], whereas receptor activation by oxidized low-density-lipoprotein-derived ligands promotes cholesterol trafficking in macrophages [17]. Recently, a somatic gene rearrangement, which results in the generation of a mutant PAX8–PPARγ fusion protein, has been reported to occur in 20–40% of human thyroid follicular tumours [4].

However, it is in relation to adipocyte biology and control of tissue insulin sensitivity that PPARγ has come to the fore, coinciding with the recent introduction of the thiazolidinediones (TZDs; e.g. rosiglitazone and pioglitazone), a novel class of anti-diabetic agent, for use in the management of Type II diabetes mellitus (T2DM). These agents are high-affinity ligands for PPARγ and reduce blood-glucose levels by enhancing tissue sensitivity to insulin action *in vivo* [18]. Recognition that PPARγ is the molecular target for these compounds strongly implicated the receptor in mammalian glucose homoeostasis and suggested that variation(s) in the human gene might be associated with alteration(s) in insulin sensitivity. Accordingly, abnormalities in glucose homoeostasis and adipogenesis have been the focus of attention for those studying aberrant PPARγ signalling in humans.

A polymorphism (Pro-12→Ala) within the unique N-terminus of PPARγ2 (allelic frequency up to 15% in certain populations), the most common human genetic variant reported to date, has been reported to afford protection against the risk of developing T2DM [19]. Indeed, it has been estimated that the global prevalence of T2DM might be reduced by as much as 25% if the entire population carried the allele containing Ala-12, thereby conferring a major population benefit for a relatively minor change in receptor function, with the less transcriptionally active Ala-12 variant promoting a lower rate of accretion of adipose tissue mass, leading to a preservation of insulin sensitivity [20]. However, there continues to be vigorous debate regarding both the extent of the proposed benefit and the mechanism through which it might occur, with recent evidence suggesting a complex interaction in which this genetic variant is influenced by environmental factors including dietary fatty acid intake [21].

A much rarer genetic mutation, Pro-115→Gln (P115Q in PPARγ2 nomenclature), which also occurs within the N-terminal domain of PPARγ, has been identified in a small number of obese subjects. *In vitro* studies suggest that the Pro-115 to Gln mutation renders the receptor more constitutively active than its WT counterpart (through disruption of receptor phosphorylation at an adjacent residue, Ser-114), and therefore predisposes to enhanced adipogenesis and obesity [22]. Moreover, preservation or even enhancement of insulin sensitivity in the face of such weight gain was also postulated as a possible component of this 'gain-of-function' phenotype, although the presence of diabetes in three of four reported subjects would seem to argue against the latter. The identification of further affected subjects with this mutation should help to clarify these observations.

Recently, we and others have reported a second class of genetic mutation in PPARγ, with the identification of loss-of-function mutations within the receptor LBD. These mutations are analogous to those previously reported in RTH, and we have therefore termed the clinical syndrome PLR.

Molecular genetics and mutant receptor properties in PLR

With the recognition that TZDs act via PPARγ, this candidate gene was screened in a cohort of 85 subjects with severe insulin resistance (defined by the co-existence of extreme hyperinsulinaemia and acanthosis nigricans). Initially we identified two different heterozygous missense mutations (Pro-467→Leu and Val-290→Met; P467L and V290M in PPARγ1 nomenclature) in the receptor LBD in three affected individuals [23]. Both mutations impaired receptor function through destabilization of helix 12, an amphipathic α-helix at the C-terminal end of the LBD which facilitates both ligand-binding and recruitment of transcriptional co-activators (Figure 2) [24]. Moreover, in a manner analogous to their TRβ counterparts in RTH, the mutant receptors inhibited WT function in a dominant-negative manner. Subsequently, other groups have identified additional loss-of-function mutations within the PPARγ LBD, Arg-425→Cys and Phe-388→Leu in

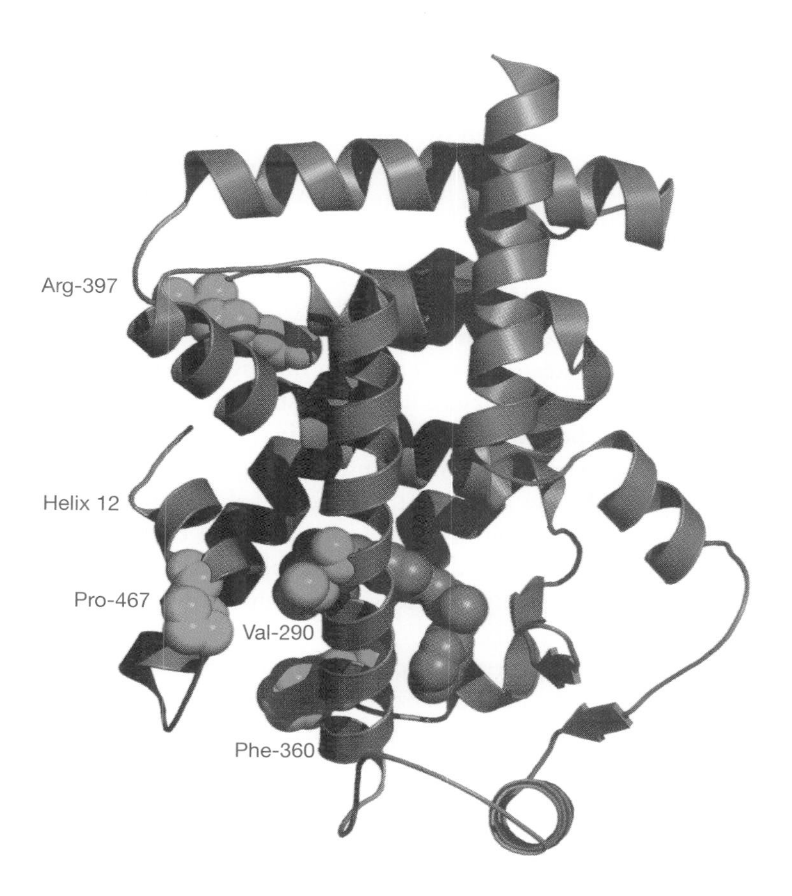

Figure 2. Loss-of-function mutations within the LBD of human PPARγ
Molecular modelling demonstrating the relative positions of the four residues (side chains shown in red with numbering corresponding to PPARγ1 nomenclature) in which missense mutations have been reported (to date) in the human PPARγ ligand-resistance syndrome. The TZD ligand rosiglitazone is shown in purple and the position of helix 12, which makes important contacts with both ligand and transcriptional co-activators, is highlighted. Note that both the Pro-467→Leu and Val-290→Met mutations are predicted to destabilize the position of helix 12, thereby markedly impairing mutant receptor function [24].

PPARγ2 (corresponding to Arg-397→Cys and Phe-360→Leu in PPARγ1; Figure 2), with all affected subjects exhibiting insulin resistance [25,26].

Clinical features and management in PLR

In keeping with the central role of PPARγ in adipogenesis, we now recognize that a stereotyped pattern of partial lipodystrophy is a cornerstone of the clinical phenotype of the PLR syndrome, with selective diminution of gluteal and limb fat but relative preservation of central adiposity. Whereas lipodystrophy itself is known to predispose to insulin resistance, detailed metabolic studies in one affected subject suggest that other factors may also contribute to its pathogenesis in this particular setting [27]. For example, subcutaneous adipocytes from a male harbouring the Pro-467→Leu mutation appear to be relatively metabolically inert, demonstrating both impaired lipogenesis and lipolysis. In addition, circulating levels of adiponectin, an adipokine that enhances insulin action in skeletal muscle and liver, were strikingly reduced in subjects harbouring PPARγ mutations when compared with healthy controls or subjects with severe insulin resistance not attributable to mutations in PPARγ [28].

Several other features of the human metabolic syndrome [including hypertension, dyslipidaemia (low high-density lipoprotein cholesterol, high triacylglycerols) and hepatic steatosis] are common findings in those with PLR, and affected females show a tendency to the polycystic ovarian syndrome. Another intriguing finding in the proband from the original Pro-467→Leu kindred was the occurrence of severe pre-eclampsia in both pregnancies. While this observation is potentially highly significant when coupled with knowledge that the murine PPARγ-gene knockout is lethal *in utero* due to combined cardiac and placental defects, its occurrence will need to be substantiated in further cases to determine whether this is truly an integral component of the human PLR syndrome (Table 3) [29].

Currently, the management of those with PLR is directed, where possible, at treatment of each individual component of the syndrome, i.e. insulin resistance/T2DM, hypertension and dyslipidaemia, together with their attendant complications. However, the availability of TZDs would appear to offer, at least in principle, a disease-specific therapy which should target the facets (lipodystrophy and insulin resistance) which underpin the syndrome. To date, we have had the opportunity to treat one individual from each original kindred with rosiglitazone. The subject from the Pro-467→Leu family exhibited a dramatic improvement in glycaemic control, which correlated with enhanced insulin sensitivity and accretion of adipose tissue. In marked contrast, the Val-290→Met proband showed very little clinical response. These findings appeared to correlate with the *in vitro* properties of the mutant receptors, with the Val-290→Met mutant exhibiting persistent dominant-negative activity even at the highest concentration of rosiglitazone, suggesting that a more potent receptor ligand might be needed to achieve a therapeutic effect [27].

Table 3. Features of the human PLR syndrome

Partial lipodystrophy (especially limb and gluteal)
Insulin resistance with or without T2DM
Dyslipidaemia (high triacylglycerols, low high-density lipoprotein cholesterol)
Hypertension
Hepatic steatosis
Polycystic ovarian syndrome
(?) Pre-eclampsia

Digenic insulin resistance

Most recently, we have studied an unusual kindred in which five subjects exhibit severe insulin resistance [30]. Screening the PPARγ gene in this family revealed a novel heterozygous frameshift/premature stop mutation within the DNA-binding domain in seven individuals. As predicted, the truncated PPARγ mutant failed to bind DNA and regulate target gene transcription. In addition, it exhibited no detectable dominant-negative activity when co-expressed with the WT receptor. Remarkably, further candidate gene studies in this family identified a second heterozygous frameshift/premature stop mutation in protein phosphatase 1 regulatory subunit 3A (PPP1R3A), which encodes a protein expressed in cardiac and skeletal muscle that regulates activity of glycogen synthase, the rate-limiting enzyme in glycogen synthesis. Functional studies with the truncated PPP1R3A mutant protein indicated that it mis-localizes intracellularly, therefore suggesting that it is unlikely to function effectively in regulating glycogen synthase activity.

Correlation of genotypes with metabolic parameters within this kindred revealed that while either gene defect alone was insufficient to mediate the clinical phenotype, double heterozygosity for both PPARγ and PPP1R3A gene defects in five subjects resulted in severe insulin resistance. To our knowledge, there is no direct link between PPARγ and PPP1R3A action and the two genes are principally expressed in different tissues (adipose tissue and skeletal muscle, respectively). This kindred represents the first reported example of a digenic cause of human insulin resistance and suggests that a 'metabolic dialogue' between fat and skeletal muscle dictates insulin sensitivity.

HNF4α and MODY1

HNF4α (NR2A1) remains a true orphan nuclear receptor with no convincing endogenous ligand(s) identified to date. Its name derives from the fact that it was first cloned from rat liver, in a manner similar to HNF1α and β, HNF3α, β and γ, and HNF6. Although the nomenclature suggests that these proteins constitute a family of hepatic transcription factors, they are structurally quite distinct and their expression extends beyond the hepatocyte with, for example, HNF1α, HNF1β and HNF4α also being expressed in pancreatic islets [31]. Intriguingly, however, and perhaps to justify the original nomenclature, the functions of several of these proteins have been shown to be intimately related; for example, in pancreatic β-cells, HNF1α, HNF1β and HNF4α regulate the expression of the insulin and other genes involved in glucose transport and metabolism. Furthermore, mutations in HNF genes are associated with a specific subtype of early onset T2DM referred to as MODY [31]. This disorder, which is characterized by defective pancreatic β-cell function, is deemed to be present when T2DM occurs in at least two generations with at least one member being affected under the age of 25 years. MODY has long been recognized to be a heterogeneous entity, and it is now clear that this is reflected in a number of different genes being involved (e.g. MODY type 1, HNF4α; type 2, glucokinase; type 3, HNF1α; MODY X, defective gene not yet identified).

In 1996, Bell and colleagues [32] were able to demonstrate linkage of MODY1 to a particular region of chromosome 20 (20q12–13) which encompasses the HNF4α gene locus [32]. Sequence analysis in a single large pedigree demonstrated that affected individuals were heterozygous for a nonsense mutation (Gln-268→Xaa) in the HNF4α gene, consistent with the dominant mode of inheritance of this disorder. To date, more than ten different missense, nonsense, frameshift and in-frame deletion/insertion mutations have been reported in the gene, predominantly in families with a MODY1 phenotype. However, in a small number of cases the possibility that the gene variation may not be responsible for MODY, but rather may represent an association with T2DM or a rare polymorphism, has not been excluded. The hyperglycaemia in subjects with HNF4α-related MODY tends to worsen with time, resulting in the need for treatment with oral hypoglycaemic agents or insulin in a significant proportion of patients [33].

It has been suggested that haploinsufficency for HNF4α function in pancreatic islet β-cells is the pathogenic basis of the MODY1 phenotype. This hypothesis is based on two observations: first, many of the diverse mutations that have been reported result in truncated proteins with major functional deficits; secondly, HNF4α exhibits high constitutive transcriptional activity and was constitutively bound to fatty acids when crystallized [34]. However, others have raised the possibility of dominant-negative activity, and the lack of a MODY phenotype in HNF4α heterozygous-null mice might provide support for this alternative hypothesis.

SHP and obesity

SHP (NR0B2) is an atypical orphan nuclear receptor which, in comparison with other nuclear receptors, lacks N-terminal and DNA-binding domains, and consists solely of a C-terminal region corresponding to a putative LBD [35]. In humans it is expressed in liver, where it has been implicated in the regulation of cholesterol and bile acid homoeostasis, and at several other sites including the pancreatic β-cells, spleen, small intestine and adrenal glands. It is capable of modulating the activity of a number of other nuclear receptors, either inhibiting (e.g. HNF4α, pregnane X receptor, LRH-1 and liver X receptor α) or augmenting (e.g. PPARγ) their transcriptional activity [36].

Due to its expression in the pancreas and ability to regulate HNF4α, SHP was considered to be a plausible candidate gene in MODY. In 2001, Nishigori and colleagues [37] reported the identification of five mutations and a single polymorphism within the gene in a cohort of 173 unrelated Japanese subjects with early-onset, non-ketotic diabetes. Interestingly, five of the six mutations that were described in this study were associated with a mildly obese phenotype rather than diabetes, with affected individuals exhibiting birth weights that were at least 1 S.D. higher than the mean when corrected for gestational age. Accordingly, the authors went on to screen a second cohort of young obese subjects (n=101; body-mass index >25 kg/m^2) and identified four SHP gene mutations, two of which differed from those found in the diabetic cohort. Combining these cohorts, 6.3% of Japanese subjects with early-onset obesity had SHP mutations. The SHP mutant proteins were impaired in their ability to inhibit HNF4α transactivation, leading the authors to speculate that such loss of SHP activity *in utero* might enhance HNF4α function leading to augmented insulin-stimulated adipogenesis. However, additional studies in other population-dependent cohorts have failed to confirm SHP mutations as a common association with obesity, although influences on birth weight have been observed [38].

SF1 and DAX1 in disorders of gonadal and adrenal development

SF1 (NR5A1) and DAX1 (NR0B1) are two orphan nuclear receptors that are expressed in the hypothalamus, pituitary gonadotrophs, gonads and adrenals and play a key role in both the development and function of the HPG axis at several levels [39].

SF1

The gene encoding SF1 is located at chromosome 9q33, and encodes a 461-amino-acid protein, which resembles other members of the nuclear receptor family. It binds to DNA as a monomer (the P-box within the first zinc-finger motif confers specificity for the hexanucleotide response element), and this interaction is stabilized through additional contacts between residues in the A-box part of the hinge region (Figure 1A) and the 5′-flanking nucleotide sequence upstream of the DNA-response element. Recently, naturally occurring SF1 mutations have been identified in two individuals with a male karyotype who exhibited complete sex reversal, with testicular dysgenesis, persistence of Mullerian structures and primary adrenal failure [40,41]. In the first subject a heterozygous missense mutation (Gly-35→Glu) was identified within the P-box of SF-1. Consistent with this, binding to and transactivation of a variety of target genes by the SF-1 mutant was impaired. In contrast, the second subject was found to be homozygous for a mutation (Arg-92→Gln) within the A-box of SF1, which weakens but does not abolish receptor binding to DNA. Heterozygous members of this second kindred were phenotypically normal. Taken together these observations suggest that, in humans, gene dosage of SF1 is critical in regulating adrenal and pituitary/gonadal axis development, with a 'modest' reduction in receptor function (as observed in the Arg-92→Gln heterozygotes) having no discernable consequences, while more 'severe' loss-of-function (as seen in the Gly-35→Glu heterozygote or the Arg-92→Gln homozygote) results in major developmental abnormalities.

A further heterozygous SF1 mutation (Arg-255→Leu) has been reported in a female (46XX) with primary adrenal failure. The presence of ovaries in this subject suggests that female gonadal development may be less dependent on SF1 function than adrenal and testis development [42].

DAX1

DAX1 is an unusual member of the nuclear receptor family, with an N-terminal domain containing tandem repeats of several Leu-Xaa-Xaa-Leu-Leu (LXXLL)-like amino acid motifs, similar to those found in transcriptional co-activators such as SRC-1. This is linked to a C-terminal putative LBD which resembles that of other nuclear receptors. Functional studies indicate that DAX1 is a transcriptional repressor, at least in part through inhibition of the activity of SF1 [43].

Human mutations in DAX1 cause X-linked AHC, a disorder of adrenal cortical development [44]. Males with this condition typically present in infancy or childhood with primary adrenal failure. Although the HPG axis appears to be intact in early life, hypogonadotrophic hypogonadism usually manifests at the time of puberty and reflects a combination of defects at both hypothalamic and pituitary levels. More than 80 different mutations have been reported in DAX1, mostly nonsense or frameshift mutations that cause premature truncation of the protein, thereby impairing its ability to function as a transcriptional repressor [39]. A small number of missense mutations have been identified in the putative LBD. As might be predicted, these mutations appear to be less deleterious, with the mutant proteins retaining some repressor activity, consistent with the milder phenotype of affected individuals. Other unusual phenotypic variants that have been reported include isolated hypogonadotrophic hypogonadism in a female homozygous for a truncation mutation in DAX1 through gene conversion [45], and delayed puberty in heterozygous female mutation carriers in one family [46]. Most recently, an unusually mild form of AHC was found in a male subject harbouring an apparently severe premature stop mutation, Gln-37→Xaa [47]. However, subsequent *in vitro* studies with this mutant revealed the unexpected generation of a partially functional N-terminally truncated DAX1 protein through the use of an alternative in-frame translation start site, thereby ameliorating the classical AHC phenotype. Taken together, these reports again emphasize the critical effects of DAX1 gene dosage in the development and function of both the HPG axis and adrenal glands.

PNR and enhanced S-cone syndrome (ESCS)

PNR (NR2E3) is a novel, recently identified orphan member of the nuclear receptor family that exhibits localized expression in retinal photoreceptor cells, suggesting that it might be involved in their differentiation and/or maintenance [48]. In support of this hypothesis, its human chromosomal location (15q24) has been independently identified as a susceptibility locus for retinal degenerative disorders [49]. Accordingly, Haider and colleagues [50] screened the PNR gene in families with Bardet–Biedl syndrome (a disorder characterized by retinitis pigmentosa, obesity, hypogonadism, renal dysfunction and mental retardation), including kindreds in whom there was clear evidence of linkage to chromosome 15, as well as others with unknown linkage [50]. No mutations in PNR were found in this original selective cohort and the authors therefore broadened their search to include a wider spectrum of retinal degenerative disease, including a small subset of individuals with the ESCS. Most hereditary human retinal degenerative disorders are associated with a reduction in the number of mature photoreceptors with consequent loss of visual function. However ESCS, an autosomal recessive retinopathy, is unusual and is characterized by a gain in photoreceptor function with affected individuals manifesting increased sensitivity to blue light mediated through the S (short-wavelength) blue cones, while simultaneously suffering visual loss (especially night blindness) to varying degrees of L (long red) and M (middle green) cone vision, and retinal degeneration. In a cohort of ESCS probands, 94% were found to harbour homozygous mutations (splice-acceptor site, deletions and missense) in PNR [50]. Although the precise mechanism through which aberrant PNR signalling leads to the phenotype of ESCS remains unclear, it has been suggested that the altered S- to L/M-cone sensitivity may reflect a return to a default pathway of cone differentiation, in which defects in PNR allow photoreceptor precursors to retain their S-cone commitment rather than switching to the L- or M-cone phenotype.

NURR1 and abnormalities of dopaminergic neurotransmission: familial Parkinson's disease and schizophrenia/manic-depressive disorder

The NURR1 (NR4A2) gene is highly conserved and encodes a member of the nerve growth factor inducible factor I-B subfamily of nuclear receptors, which plays a critical role in mesencephalic dopamine neuronal development. Accordingly, NURR1 has been implicated as a potential candidate gene in conditions as diverse as Parkinson's disease and schizophrenia/manic-depressive disorder.

Recently two nucleotide changes (-291Tdel and -245T to G), which map to the first non-coding exon of NURR1 (upstream of the transcriptional start site), were found to affect one in ten alleles from 107 individuals with familial Parkinson's disease, but not subjects with sporadic Parkinson's disease or unaffected controls [51]. *In vitro* studies demonstrated a dramatic reduction in mutant receptor mRNA compared with WT NURR1 transcript in cells transfected with expression vectors containing either the mutant or WT alleles, which correlated with a significant attenuation of receptor-mediated transactivation from a NuRE containing reporter gene in the presence of either mutant. Furthermore, expression of tyrosine hydroxylase, an enzyme mediating dopamine biosynthesis, was markedly reduced in cells transfected with the mutant NURR1 allele. Studies in lymphocytes from two individuals with the -291Tdel mutation demonstrated a significant reduction in NURR1 mRNA to levels that were even lower than 50% (the level of transcripts that might be expected from the remaining WT allele), leading the authors to speculate that the mutant might inhibit expression of its WT counterpart in a dominant-negative manner. Further studies are awaited to confirm these findings and to determine the mechanisms through which aberrant NURR1 signalling induces the Parkinson's disease phenotype.

In contrast with Parkinson's disease, the precise role of the dopaminergic system in the pathogenesis of schizophrenia remains unclear, although the effectiveness of anti-dopaminergic agents therapeutically strongly implicates genes involved in dopamine cell development and maintenance. To date, only a handful of mutations in NURR1 (in exons 1and 3) have been found in patients with schizophrenia and manic-depressive disorder, and larger-cohort studies are awaited to validate these findings [52,53]. In addition, it will be a challenge to explain how apparent loss-of-function mutations within NURR1 can be associated with disorders as divergent as Parkinson's disease and schizophrenia/manic-depressive disorder, which are treated by enhancing or blocking dopaminergic neurotransmission respectively.

Conclusions

The study of human subjects harbouring naturally occurring mutations in various members of the steroid nuclear receptor superfamily has greatly enhanced our understanding of the diverse roles that these receptors play in normal mammalian physiology. Often these observations have complemented those derived from animal models, and in many cases they have served to emphasize the important differences that exist between the species. However, for a significant number of nuclear receptors that are well characterized, the human disorders or phenotypes associated with receptor defects remain to be elucidated. It is important that we continue to search for these 'human receptor disorders', not only to study the pathogenesis of the phenotype in affected individuals, but also to verify that the conclusions that have been drawn currently regarding the roles of such receptors in human physiology are correct.

Summary

- *Mutations in human TRβ are associated with the syndrome of RTH, a disorder with a distinctive biochemical signature of elevated free thyroid hormones with an unsuppressed thyroid-stimulating hormone.*
- *Mutations in the LBD of human PPARγ are associated with a novel syndrome (PLR), which is characterized by a stereotyped pattern of partial lipodystrophy, insulin resistance, hypertension and other features of the human metabolic syndrome.*

- *Mutations in various orphan nuclear receptors are associated with a diverse group of human disorders involving aspects of metabolism (HNF4α, SHP), gonadal and adrenal development (SF1, DAX1), retinal function (PNR) and neuronal dopaminergic transmission (NURR1).*

We thank J.W.R. Schwabe for undertaking the molecular modelling shown in Figure 2 and T.D. Wallman for secretarial assistance.

References

1. Evans, R.M. (1988) The steroid and thyroid hormone receptor superfamily. *Science* **240**, 889–895
2. Kliewer, S.A., Umesono, K., Mangelsdorf, D.J. & Evans, R.M. (1992) Retinoid X receptor interacts with nuclear receptors in retinoic acid, thyroid hormone and vitamin D3 signalling. *Nature (London)* **355**, 446–449
3. Lin, J.R., Nagy, L., Satoshi, I., Shao, W., Miller, W. & Evans, R.M. (1998) Role of the histone deacetylase complex in acute promyelocytic leukaemia. *Nature (London)* **391**, 811–814
4. Kroll, T.G., Sarraf, P., Pecciarini, L., Chen, C.J., Mueller, E., Spiegelman, B.M. & Fletcher, J.A. (2000) PAX8-PPARγ1 fusion oncogene in human thyroid carcinoma. *Science* **289**, 1357–1360
5. Refetoff, S., De Wind, L.T. & De Groot, L.J. (1967) Familial syndrome combining deaf-mutism, stippled epiphyses, goiter and abnormally high PBI: possible target organ refractoriness to thyroid hormone. *J. Clin. Endocrinol. Metab.* **27**, 279–294
6. Refetoff, S., Weiss, R.E. & Usala, S.J. (1993) The syndromes of resistance to thyroid hormone. *Endocr. Rev.* **14**, 348–399
7. Usala, S.J., Bale, A.E., Gesundheit, N., Weinberger, C., Lash, R.W., Wondisford, F.E., McBride, O.W. & Weintraub, B.D. (1988) Tight linkage between the syndrome of generalized thyroid hormone resistance and the human c-erbA β gene. *Mol. Endocrinol.* **2**, 1217–1220
8. Weiss, R.E., Xu, J., Ning, G., Pohlenz, J., O'Malley, B.W. & Refetoff, S. (1999) Mice deficient in the steroid receptor coactivator 1 (SRC-1) are resistant to thyroid hormone. *EMBO J.* **18**, 1900–1904
9. Weiss, R.E., Gehin, M., Xu, J., Sadow, P.M., O'Malley, B.W., Chambon, P. & Refetoff, S. (2002) Thyroid function in mice with compound heterozygous and homozygous disruptions of SRC-1 and TIF-2 coactivators: evidence for haploinsufficiency. *Endocrinology* **143**, 1554–1557
10. Chatterjee, V.K.K., Nagaya, T., Madison, L.D., Datta, S., Rentoumis, A. & Jameson, J.L. (1991) Thyroid hormone resistance syndrome: inhibition of normal receptor function by mutant thyroid receptors. *J. Clin. Invest.* **87**, 1977–1984
11. Forrest, D., Hanebuth, E., Smeyne, R.J., Everds, N., Stewart, C.L., Wehner, J.M. & Curran, T. (1996) Recessive resistance to thyroid hormone in mice lacking thyroid hormone receptor β: evidence for tissue-specific modulation of receptor function. *EMBO J.* **15**, 3006–3015
12. Ono, S., Schwartz, I.D., Mueller, O.T., Root, A.W., Usala, S.J. & Bercu, B.B. (1991) Homozygosity for a dominant negative thyroid hormone receptor gene responsible for generalized resistance to thyroid hormone. *J. Clin. Endocrinol. Metab.* **73**, 990–994
13. Adams, M., Matthews, C.H., Collingwood, T.N., Tone, Y., Beck Peccoz, P. & Chatterjee, V.K.K. (1994) Genetic analysis of twenty-nine kindreds with generalised and pituitary resistance to thyroid hormone. *J. Clin. Invest.* **94**, 506–515
14. Gurnell, M., Rajanayagam, O., Barbar, I., Keston Jones, M. & Chatterjee, V.K.K. (1998) Reversible pituitary enlargement in the syndrome of resistance to thyroid hormone. *Thyroid* **8**, 679–682
15. Tontonoz, P., Hu, E. & Spiegelman, B.M. (1994) Stimulation of adipogenesis in fibroblasts by PPARγ2, a lipid-activated transcription factor. *Cell* **79**, 1147–1156
16. Ricote, M., Li, A.C., Willson, T.M., Kelly, C.J. & Glass, C.K. (1998) The peroxisome proliferator-activated receptor-γ is a negative regulator of macrophage activation. *Nature (London)* **391**, 79–86

17. Lee, C.H. & Evans, R.M. (2002) Peroxisome proliferator-activated receptor-γ in macrophage lipid homeostasis. *Trends Endocrinol. Metab.* **13**, 331–335
18. Henry, R.R. (1997) Thiazolidinediones. *Endocrinol. Metab. Clin. North Am.* **26**, 553–573
19. Altshuler, D., Hirschhorn, J.N., Klannemark, M., Lindgren, C.M., Vohl, M.C., Nemesh, J., Lane, C.R., Schaffner, S.F., Bolk, S., Brewer, C. et al. (2000) The common PPARγ Pro12Ala polymorphism is associated with decreased risk of type 2 diabetes. *Nat. Genet.* **26**, 76–80.
20. Deeb, S.S., Fajas, L., Nemoto, M., Pihlajamaki, J., Mykkanen, L., Kuusisto, J., Laakso, M., Fujimoto, W. & Auwerx, J. (1998) A Pro12Ala substitution in PPARγ2 associated with decreased receptor activity, lower body mass index and improved insulin sensitivity. *Nat. Genet.* **20**, 284–287
21. Luan, J., Browne, P.O., Harding, A.H., Halsall, D.J., O'Rahilly, S., Chatterjee, V.K. & Wareham, N.J. (2001) Evidence for gene-nutrient interaction at the PPARγ locus. *Diabetes* **50**, 686–689
22. Ristow, M., Muller-Wieland, D., Pfeiffer, A., Krone, W. & Kahn, C.R. (1998) Obesity associated with a mutation in a genetic regulator of adipocyte differentiation. *N. Engl. J. Med.* **339**, 953–959
23. Barroso, I., Gurnell, M., Crowley, V.E.F., Agostini, M., Schwabe, J.W., Soos, M.A., Maslen, G.L.I., Williams, T.D.M., Lewis, H., Schafer, A.J. et al. (1999) Dominant negative mutations in human PPARγ are associated with severe insulin resistance, diabetes mellitus and hypertension. *Nature (London)* **402**, 880–883
24. Kallenberger, B.C., Love, J.D., Chatterjee, V.K. & Schwabe, J.W. (2003) A dynamic mechanism of nuclear receptor activation and its perturbation in a human disease. *Nat. Struct. Biol.* **10**, 136–140
25. Agarwal, A.K. & Garg, A. (2002) A novel heterozygous mutation in peroxisome proliferator-activated receptor-γ gene in a patient with familial partial lipodystrophy. *J. Clin. Endocrinol. Metab.* **87**, 408–411
26. Hegele, R.A., Cao, H., Frankowski, C., Mathews, S.T. & Leff, T. (2002) PPARG F388L, a transactivation-deficient mutant, in familial partial lipodystrophy. *Diabetes* **51**, 3586–3590
27. Savage, D.B., Tan, G.D., Acerini, C.L., Jebb, S.A., Agostini, M., Gurnell, M., Williams, R., Umpleby, A.M., Thomas, E.L., Bell, J.D. et al. (2003) Clinical and pathophysiological features of a metabolic syndrome resulting from mutations in the nuclear receptor PPARγ. *Diabetes* **52**, 910–917
28. Combs, T.P., Wagner, J.A., Berger, J., Doebber, T., Wang, W.J., Zhang, B.B., Tanen, M., Berg, A.H., O'Rahilly, S., Savage, D.B. et al. (2002) Induction of adipocyte complement-related protein of 30 kilodaltons by PPARγ a agonists: a potential mechanism of insulin sensitization. *Endocrinology* **143**, 998–1007
29. Gurnell, M., Savage, D.B., Chatterjee, V.K. & O'Rahilly, S. (2003) The metabolic syndrome: peroxisome proliferator-activated receptorγ and its therapeutic modulation. *J. Clin. Endocrinol. Metab.* **88**, 2412–2421
30. Savage, D.B., Agostini, M., Barroso, I., Gurnell, M., Luan, J., Meirhaeghe, A., Harding, A.H., Ihrke, G., Rajanayagam, O., Soos, M.A. et al. (2002) Digenic inheritance of severe insulin resistance in a human pedigree. *Nat. Genet.* **31**, 379–384
31. Ryffel, G.U. (2001) Mutations in the human genes encoding the transcription factors of the hepatocyte nuclear factor (HNF)1 and HNF4 families: functional and pathological consequences. *J. Mol. Endocrinol.* **27**, 11–29
32. Yamagata, K., Furuta, H., Oda, N., Kaisaki, P.J., Menzel, S., Cox, N.J., Fajans, S.S., Signorini, S., Stoffel, M. & Bell, G.I. (1996) Mutations in hepatocyte nuclear factor-4α gene in maturity-onset diabetes of the young (MODY1). *Nature (London)* **384**, 458–460
33. Fajans, S.S., Bell, G.I. & Polonsky, K.S. (2001) Molecular mechanisms and clinical pathophysiology of maturity-onset diabetes of the young. *N. Engl. J. Med.* **345**, 971–980
34. Wisely, G.B., Miller, A.B., Davis, R.G., Thornquest, Jr, A.D., Johnson, R., Spitzer, T., Sefler, A., Shearer, B., Moore, J.T., Willson, T.M. & Williams, S.P. (2002) Hepatocyte nuclear factor 4 is a transcription factor that constitutively binds fatty acids. *Structure (Cambridge)* **10**, 1225–1234
35. Seol, W., Choi, H.S. & Moore, D.D. (1996) An orphan nuclear hormone receptor that lacks a DNA binding domain and heterodimerizes with other receptors. *Science* **272**, 1336–1339
36. Nishizawa, H., Yamagata, K., Shimomura, I., Takahashi, M., Kuriyama, H., Kishida, K., Hotta, K., Nagaretani, H., Maeda, N., Matsuda, M. et al. (2002) Small heterodimer partner, an orphan

nuclear receptor, augments peroxisome proliferator-activated receptorγ transactivation. *J. Biol. Chem.* **277**, 1586–1592

37. Nishigori, H., Tomura, H., Tonooka, N., Kanamori, M., Yamada, S., Sho, K., Inoue, I., Kikuchi, N., Onigata, K., Kojima, I. et al. (2001) Mutations in the small heterodimer partner gene are associated with mild obesity in Japanese subjects. *Proc. Natl. Acad. Sci. U.S.A.* **98**, 575–580
38. Hung, C.C., Farooqi, I.S., Ong, K., Luan, J., Keogh, J.M., Pembrey, M., Yeo, G.S., Dunger, D., Wareham, N.J. & O'Rahilly, S. (2003) Contribution of variants in the small heterodimer partner gene to birthweight, adiposity, and insulin levels: mutational analysis and association studies in multiple populations. *Diabetes* **52**, 1288–1291
39. Achermann, J.C., Ozisik, G., Meeks, J.J. & Jameson, J.L. (2002) Genetic causes of human reproductive disease. *J. Clin. Endocrinol. Metab.* **87**, 2447–2454
40. Achermann, J.C., Ito, M., Hindmarsh, P.C. & Jameson, J.L. (1999) A mutation in the gene encoding steroidogenic factor-1 causes XY sex reversal and adrenal failure in humans. *Nat. Genet.* **22**, 125–126
41. Ito, M., Achermann, J.C. & Jameson, J.L. (2000) A naturally occurring steroidogenic factor-1 mutation exhibits differential binding and activation of target genes. *J. Biol. Chem.* **275**, 31708–31714
42. Biason-Lauber, A. & Schoenle, E.J. (2000) Apparently normal ovarian differentiation in a prepubertal girl with transcriptionally inactive steroidogenic factor 1 (NR5A1/SF-1) and adrenocortical insufficiency. *Am. J. Hum. Genet.* **67**, 1563–1568
43. Ito, M., Yu, R. & Jameson, J.L. (1997) DAX-1 inhibits SF-1-mediated transactivation via a carboxy-terminal domain that is deleted in adrenal hypoplasia congenita. *Mol. Cell. Biol.* **17**, 1476–1483
44. Muscatelli, F., Strom, T.M., Walker, A.P., Zanaria, E., Recan, D., Meindl, A., Bardoni, B., Guioli, S., Zehetner, G., Rabl, W. et al. (1994) Mutations in the DAX-1 gene give rise to both X-linked adrenal hypoplasia congenita and hypogonadotropic hypogonadism. *Nature (London)* **372**, 672–676
45. Merke, D.P., Tajima, T., Baron, J. & Cutler, G.B., Jr. (1999) Hypogonadotropic hypogonadism in a female caused by an X-linked recessive mutation in the DAX1 gene. *N. Engl. J. Med.* **340**, 1248–1252
46. Seminara, S.B., Achermann, J.C., Genel, M., Jameson, J.L. & Crowley, Jr, W.F. (1999) X-linked adrenal hypoplasia congenita: a mutation in DAX1 expands the phenotypic spectrum in males and females. *J. Clin. Endocrinol. Metab.* **84**, 4501–4509
47. Ozisik, G., Mantovani, G., Achermann, J.C., Persani, L., Spada, A., Weiss, J., Beck-Peccoz, P. & Jameson, J.L. (2003) An alternate translation initiation site circumvents an amino-terminal DAX1 nonsense mutation leading to a mild form of X-linked adrenal hypoplasia congenita. *J. Clin. Endocrinol. Metab.* **88**, 417–423
48. Kobayashi, M., Takezawa, S., Hara, K., Yu, R.T., Umesono, Y., Agata, K., Taniwaki, M., Yasuda, K. & Umesono, K. (1999) Identification of a photoreceptor cell-specific nuclear receptor. *Proc. Natl. Acad. Sci. U.S.A.* **96**, 4814–4819
49. Carmi, R., Elbedour, K., Stone, E.M. & Sheffield, V.C. (1995) Phenotypic differences among patients with Bardet–Biedl syndrome linked to three different chromosome loci. *Am. J. Med. Genet.* **59**, 199–203
50. Haider, N.B., Jacobson, S.G., Cideciyan, A.V., Swiderski, R., Streb, L.M., Searby, C., Beck, G., Hockey, R., Hanna, D.B., Gorman, S. et al. (2000) Mutation of a nuclear receptor gene, NR2E3, causes enhanced S cone syndrome, a disorder of retinal cell fate. *Nat. Genet.* **24**, 127–131
51. Le, W.D., Xu, P., Jankovic, J., Jiang, H., Appel, S.H., Smith, R.G. & Vassilatis, D.K. (2003) Mutations in NR4A2 associated with familial Parkinson disease. *Nat. Genet.* **33**, 85–89
52. Buervenich, S., Carmine, A., Arvidsson, M., Xiang, F., Zhang, Z., Sydow, O., Jonsson, E.G., Sedvall, G.C., Leonard, S., Ross, R.G. et al. (2000) NURR1 mutations in cases of schizophrenia and manic-depressive disorder. *Am. J. Med. Genet.* **96**, 808–813
53. Chen, Y.H., Tsai, M.T., Shaw, C.K. & Chen, C.H. (2001) Mutation analysis of the human NR4A2 gene, an essential gene for midbrain dopaminergic neurogenesis, in schizophrenic patients. *Am. J. Med. Genet.* **105**, 753–757

Subject index

A

A-box, 170, 172, 183
acetylation, 3, 5–6, 74, 79, 80, 81, 84–85, 94, 95
activating mutation, 152
activation function, 3, 29, 31–33, 34, 35, 36–37, 68, 109, 110, 113, 124, 125, 170
activator of heat-shock protein 90 ATPase, 49
activator protein 1, 99, 101
acute promyelocytic leukaemia, 171
adrenal hypoplasia congenita, 171, 184
AF (*see* activation function)
Aha (*see* activator of heat-shock protein 90 ATPase)
AHC (*see* adrenal hypoplasia congenita)
AIS (*see* androgen-insensitivity syndrome)
aldosterone (*see* mineralocorticoid)
Alien, 91, 93, 101
allosteric effects of hormone response elements, 68
amiloride-sensitive sodium channel, 139, 142, 149
androgen, 6–8, 12, 67–68, 114, 122, 125–134, 143, 147–149, 160, 161, 164
androgen receptor, 4, 6, 7, 8, 60, 61, 67–68, 69, 107, 117, 122–134
androgen-insensitivity syndrome, 125–130
anti-hormone, 96–97
AP-1 (*see* activator protein 1)
AR (*see* androgen receptor)
ASSC (*see* amiloride-sensitive sodium channel)

B

Bcl-2-associated gene product-1 (BAG-1), 49
breast cancer, 96–97, 130–132, 158, 159, 162, 164
bulimia, 163, 164

C

Caenorhabditis elegans, 20–21
cancer, 96–97, 130–132, 133–134, 158, 159, 160–161, 163, 164
CAR (*see* constitutive androstane receptor)
cardiovascular disease, 161, 164
caveolin-1, 115
central nervous system disease, 162, 164
chicken ovalbumin upstream promoter-transcription factor (*see* CUOP-TF)
ChIP (*see* chromatin immunoprecipitation assay)
CHIP (*see* C-terminus of heat-shock-protein-70-interacting protein)
chromatin, 6, 7, 30, 74, 75, 79, 80–81, 91–92, 94–96, 101
chromatin architecture, 79, 82, 83
chromatin immunoprecipitation assay, 7, 81, 83
chromatin remodelling complex, 78, 79, 81–83, 86
co-activator, 4, 7, 8, 17, 30, 106, 113, 117, 141, 145, 171, 172–173, 175, 178, 179, 184
(*see also* nuclear receptor co-activator; steroid receptor co-activator)
codon 200 mutation, 175

codon 300 mutation, 175
codon 400 mutation, 175
comparative genomics, 13, 16, 20
constitutive androstane receptor, 4, 13
co-regulator, 36, 124, 125, 126, 129
co-repressor, 4, 7, 8, 17, 30, 90–94, 95, 96–98, 100, 101, 171, 172–173, 176 (*see also* ligand-dependent nuclear receptor co-repressor; nuclear receptor co-repressor; small unique nuclear receptor co-repressor)
corticotropin-releasing hormone, 142, 147, 148
cortisol (*see* glucocorticoid)
COUP-TF, 13, 16
CRC (*see* chromatin remodelling complex)
CRH (*see* corticotropin-releasing hormone)
crosstalk, 90, 98–99, 101
CTE (*see* C-terminal extension)
C-terminal extension, 3, 66–67, 68, 69
C-terminus of heat-shock-protein-70-interacting protein, 53, 55
CVD (*see* cardiovascular disease)
cyclophilin, 50–51
    CyP40, 51

D

DAX1, 93, 101, 174, 183, 184
DBD (*see* DNA-binding domain)
deuterostome, 14, 15
direct repeat, 60, 62, 66, 67, 68
DNA-binding domain, 2–5, 7, 8, 12, 16, 28, 29, 33–34, 60, 62–69, 106, 107, 122–125, 127–128, 139–140, 145, 158, 170, 172–173
DNA-recognition helix, 3, 33, 63
Drosophila, 20–21, 23
dynein, 50–52, 56

E

ecdysone, 12, 13, 23
ecdysone receptor (EcR), 13
EGF (*see* epidermal growth factor)
enhanced S-cone syndrome, 174, 185–185
epidermal growth factor, 114
ER (*see* oestrogen receptor)
ERR (*see* oestrogen-related receptor)
ESCS (*see* enhanced S-cone syndrome)
everted repeat, 62

F

familial Parkinson's disease, 185–186
farnesoid X receptor (FXR), 21
FK506-binding protein (FKBP), 50–51
    FKBP51, 143
    FKBP52, 51
fluorescence recovery after photo-bleaching (FRAP), 7
fushi tarazu transcription factor 1 (FTZ-F1), 13, 16

G

gain-of-function disease, 130–132
geldanamycin, 42, 45, 50, 52, 53, 55
Gene Mutation database, 124, 126
generalized resistance to thyroid hormone, 7, 175, 176
gene silencing, 90–96, 100–101
glucocorticoid, 6, 7, 138, 139, 142, 144, 147, 148, 152
glucocorticoid receptor, 2, 4–7, 31, 32, 34, 37, 42–56, 60, 61, 63–69, 138–152
glucocorticoid resistance, 138, 143–145, 147, 148
glucocorticoid response element, 139, 142, 145
G-protein-coupled receptor, 110–112, 114
GPCR (*see* G-protein-coupled receptor)
GR (*see* glucocorticoid receptor)
GRE (*see* glucocorticoid response element)
GRTH (*see* generalized resistance to thyroid hormone)
guinea pig, 143

H

Hairless, 91, 93
HDAC (*see* histone deacetylase)

heat-shock protein 70, 42, 43, 45–49, 52, 53, 55, 56
heat-shock protein 70-interacting protein, 49
heat-shock protein 90, 41–56
heat-shock protein organizer protein, 45–47, 53
helix 12, 90, 92, 97, 100
hepatocyte nuclear factor 4α, 13, 16, 19, 171, 174, 181–183
Hip (*see* heat-shock protein 70-interacting protein)
histone acetyltransferase, 79, 141
histone code, 79, 80
histone deacetylase, 91–96, 98, 171
histone kinase, 79
histone methyltransferase, 79, 94, 95
histone modification, 74, 79–83
histone-modifying enzyme, 79, 81
HNF4α (*see* hepatocyte nuclear factor 4α)
Hop (*see* heat-shock protein organizer protein)
hormone resistance syndrome, 7, 8 (*see also* androgen-insensitivity syndrome; generalized resistance to thyroid hormone; glucocrticoid resistance; mineralocorticoid resistance; pituitary resistance to thyroid hormone; PPARγ ligand resistance; thyroid hormone resistance syndrome)
hormone-response element, 4, 12, 33–34, 60, 61–63, 66, 68, 69
HPA axis (*see* hypothalamic/pituitary/adrenal axis)
HRE (*see* hormone-response element)
hsp70 (*see* heat-shock protein 70)
hsp90 (*see* heat-shock protein 90)
hypothalamic/pituitary/adrenal axis, 138, 142, 148

I

immune system disease, 162
immunophilin, 47, 50–52, 56
infertility, 133
insulin resistance, 171, 178–179, 180, 181
inverted repeat, 60, 61, 62, 67

K

Kennedy disease (*see* spinobulbar muscular atrophy)

L

LBD (*see* ligand-binding domain)
LCoR (*see* ligand-dependent nuclear receptor co-repressor)
ligand-activated nuclear receptor (*see* constitutive androstane receptor; farnesoid X receptor; liver X receptor α; peroxisome-proliferator-activated receptor γ; pregnane X receptor; retinoic acid receptor; retinoid X receptor; thyroid hormone receptor; vitamin D receptor)
ligand-binding domain, 2–5, 7, 8, 12, 16, 18, 19, 20, 28, 31, 34–35, 36, 37, 42, 106, 113, 115, 122–125, 126, 127–130, 139–140, 145, 150, 152, 158, 170, 172–173, 175, 178, 182, 184
ligand-binding pocket, 5, 143, 145, 150, 152
ligand-dependent nuclear receptor co-repressor, 91, 93
lipodystrophy, 171, 179–180
liver X receptor α, 5, 182
loss-of-function disease, 127–128
loss-of-function mutation, 138, 147, 148, 149
LXRα (*see* liver X receptor α)
LXXLL motif, 34, 77, 91, 141

M

male infertility, 133
MAPK (*see* mitogen-activated protein kinase)
maturity onset diabetes of the young type 1, 171, 174, 181–182
mediator, 7, 76, 77, 80, 82, 83
metazoa, 11, 13–16
mineralocorticoid, 138, 139, 141, 142, 143, 148, 152

mineralocorticoid receptor, 4, 60, 61, 138–152
mineralocorticoid resistance, 138
mitogen-activated protein kinase, 6, 107, 108–109, 111–113, 116
MNAR (*see* modulation of non-genomic activity of oestrogen receptor)
modulation of non-genomic activity of oestrogen receptor, 110, 113, 117
MODY1 (*see* maturity onset diabetes of the young type 1)
molecular chaperone, 4, 6, 41–56, 106
MR (*see* mineralocorticoid receptor)

N

NCoR (*see* nuclear receptor co-repressor)
negative response element, 7, 97–98, 101
nerve growth factor inducible factor I-B (NGFI-B), 4, 5, 8, 13, 18, 60
New World monkey, 143
NF-κB (*see* nuclear factor κB)
NGFI-B (*see* nerve growth factor inducible factor I-B),
nitric oxide synthase, 44, 112, 114, 116
NR box (*see* nuclear receptor box)
NTD (*see* N-terminal transactivation domain)
N-terminal transactivation domain, 3, 28–33, 34, 36, 37
nuclear factor κB, 99, 141
nuclear receptor
  evolution , 2, 13–23
  phylogeny, 13–18
nuclear receptor box, 77, 80
nuclear receptor co-activator, 17, 73, 76–79, 84
nuclear receptor co-repressor, 17, 91–92, 93, 95, 96, 100, 171
nucleosome remodelling, 94, 95
NURR1 (*see* NUR-related factor 1)
NUR-related factor 1, 18, 173, 174, 185–186

O

obesity, 178, 182, 183, 185
oestradiol (*see* oestrogen)
oestrogen, 6, 12, 18, 107, 112–117, 133, 134, 143, 158–162
oestrogen receptor, 2, 4, 6–8, 60, 61, 63–69, 107, 108, 110, 112–117
  oestrogen receptor α (ERα), 158–164
  oestrogen receptor β (ERβ), 158–164
  oestrogen-receptor-α-knock-out (αERKO), 158–162
  oestrogen-receptor-β-knock-out (βERKO), 158–162
oestrogen-related receptor (ERR), 13, 20
orphan nuclear receptor, 4, 5, 8, 12, 18, 170, 171 (*see also* COUP-TF; DAX1; ERR; FTZ-F1; NGFI-B; NURR1; RORα; SF1; SHP)
osteoporosis, 161, 162, 163–164

P

p23, 45, 47–49, 52
p300, 77–81, 83–85
p300/CBP-associated factor (pCAF), 17, 18, 69, 78–81, 83, 141, 171
PAX8–PPARγ fusion protein, 171, 177
P-box, 170, 183
pCAF (*see* p300/CBP-associated factor)
peptidylprolyl isomerase, 50, 51, 56
peroxisome-proliferator-activated receptor γ, 4, 5, 170, 171, 172–173, 174, 177, 178–180, 181
PHA1 (*see* pseudohypoaldosteronism type 1)
phosphoinositide 3-kinase/Akt signalling pathway, 108, 112, 114, 116
phosphorylation, 3, 5–6, 32, 74, 79, 81, 85, 94, 95
photoreceptor-specific nuclear receptor, 172–173, 174, 184–185
PIC (*see* pre-initiation complex)
PI 3-kinase (*see* phosphoinositide 3-kinase/Akt signalling pathway)
pituitary resistance to thyroid hormone, 176–177
PLR (*see* PPARγ ligand resistance)
PNR (*see* photoreceptor-specific nuclear receptor)

polyglutamine repeat, 124, 131, 133
polymorphism, 162–164
polypyrimidine tract binding protein-associated splicing factor, 91, 93
post-translational modification (*see* acetylation; phosphorylation; sumoylation/SUMO-1; ubiquitination)
PP5, 51
PPARγ (*see* peroxisome-proliferator-activated receptor γ)
PPARγ ligand resistance (PLR), 177, 180
PPlase (*see* peptidylprolyl isomerase)
PPP1R3A (*see* protein phosphatase 1 regulatory subunit 3A)
PR (*see* progesterone receptor)
prairie vole, 143–144
pre-adipocyte differentiation, 177
pregnane X receptor, 182
pre-initiation complex, 74–75, 83
PRMT (*see* protein arginine methyltransferase)
progesterone, 6, 12, 67–112, 116–117, 143, 152, 161
progesterone receptor, 2, 4, 6, 60, 61, 63, 107, 108–112, 116
progestin (*see* progesterone)
prostate cancer, 96, 130, 132, 133–134, 160–161, 163, 164
prostome, 14, 15
proteasome, 43, 44, 53–55, 85
protein arginine methyltransferase, 77, 79, 81
protein–DNA contact, 63, 65, 67, 69
protein phosphatase 1 regulatory subunit 3A, 181
protein trafficking, 42, 50–52
PRTH (*see* pituitary resistance to thyroid hormone)
pseudohypoaldosteronism type 1, 138, 149–152
PSF (*see* polypyrimidine tract binding protein-associated splicing factor)
PXR (*see* pregnane X receptor)

R

RAR (*see* retinoic acid receptor)
receptor-interacting protein, 91, 93, 140
receptor trafficking, 6, 50–52
receptor turnover, 53–55, 84, 85
response element, 170, 171, 173, 183
retinoic acid receptor, 2, 3, 4, 13, 15, 18, 21, 43–44, 171, 172
retinoid-related orphan receptor, 4, 13, 18
retinoid X receptor, 3, 4, 34, 60, 62, 66, 171, 173, 176
RIP140 (*see* receptor-interacting protein 140)
RNA polymerase II, 74–77, 84
RORα (*see* retinoid-related orphan receptor)
RXR (*see* retinoid X receptor)

S

SBMA (*see* spinobulbar muscular atrophy)
schizophrenia/manic depressive disorder, 185–186
sex-hormone-binding globulin, 114
SF1 (*see* steroidogenic factor 1)
SHBG (*see* sex-hormone-binding globulin)
SHP (*see* small heterodimer partner)
silencing mediator of repressed transcription (*see* SMRTER)
silencing mediator of retinoic acid and thyroid hormone receptor (*see* SMRT)
single-nucleotide polymorphism, 163–164
small heterodimer partner, 171, 174, 182, 183
small unique nuclear receptor co-repressor, 91, 93
SMRT, 91–92, 93, 95, 96, 98, 100, 171
SMRTER, 91–92
SNP (*see* single-nucleotide polymorphism)
spinobulbar muscular atrophy, 130–131
sponge, 16
SRC (*see* steroid receptor co-activator)
Src tyrosine kinase, 6, 107, 108–109, 111, 112, 113, 116, 117

steroid hormone, 12, 13, 18, 23
(*see also* androgen; ecdysone; glucocorticoid; oestrogen; progesterone)
steroidogenic factor 1, 4, 13, 19, 171, 174, 183
steroid receptor (*see* androgen receptor; ecdysone receptor; glucocorticoid receptor; mineralocorticoid receptor; oestrogen receptor; progesterone receptor)
steroid receptor co-activator, 77, 83, 85, 86, 141, 171, 175
sumoylation/SUMO-1, 3, 5–6, 33, 85
SUN-CoR (*see* small unique nuclear receptor co-repressor)
SWI/SNF, 18, 76, 77, 78, 79, 86

T

TAF (*see* TATA-box-binding-protein-associated factor)
TATA-box-binding protein (TBP), 7, 17, 75, 76, 77, 96
TATA-box-binding-protein-associated factor, 75–76, 78
TBP (*see* TATA-box-binding protein)
testosterone (*see* androgen)
thiazolidinedione (TZD), 177–179, 180
thyroid hormone receptor, 2–5, 7, 8, 66
thyroid hormone receptor α (TRα), 173, 175
thyroid hormone receptor β (TRβ), 173, 174, 175–177, 178
thyroid hormone resistance syndrome (*see* generalized resistance to thyroid hormone; pituitary resistance to thyroid hormone)
TR (*see* thyroid hormone receptor)
transactivation domain (*see* activation function)
transciptional synergism, 81, 83
transcription factor (*see also* activator protein 1; nuclear factor κB)
basal, 7, 74, 75, 91, 94, 96, 101
transcription co-activator (*see* co-activator)
transcription co-repressor (*see* co-repressor)
transcription reinitiation, 83
transgenic animal
nuclear receptor knock-out, 8, 142, 158–162
TZD (*see* thiazolidinedione)

U

ubiquitin, 43, 44, 53–55, 77, 85
ubiquitination, 3, 5–6, 53, 55

V

vascular protection by oestrogen, 112, 116
VDR (*see* vitamin D receptor)
v-ErbA oncogene, 100
vitamin D receptor, 3–5, 7, 13, 18, 60–61, 68, 174

X

Xenopus oocyte maturation, 107, 108, 110–111

Z

zinc finger, 33, 122, 124, 170, 172–173